Study and Communication Skills for the BIOSCIENCES

Study and Communication Skills for the

BIOSCIENCES

Stuart Johnson and Jon Scott

OXFORD
UNIVERSITY PRESS

OXFORD

UNIVERSITY PRESS

Great Clarendon Street, Oxford OX2 6DP

Oxford University Press is a department of the University of Oxford.
It furthers the University's objective of excellence in research, scholarship,
and education by publishing worldwide in

Oxford New York

Auckland Cape Town Dar es Salaam Hong Kong Karachi
Kuala Lumpur Madrid Melbourne Mexico City Nairobi
New Delhi Shanghai Taipei Toronto

With offices in

Argentina Austria Brazil Chile Czech Republic France Greece
Guatemala Hungary Italy Japan Poland Portugal Singapore
South Korea Switzerland Thailand Turkey Ukraine Vietnam

Oxford is a registered trade mark of Oxford University Press
in the UK and in certain other countries

Published in the United States
by Oxford University Press Inc., New York

British Library Cataloguing in Publication Data

Data available

Library of Congress Cataloging-in-Publication Data

Johnson, Stuart.
Study and communication skills for the biosciences / Stuart Johnson and Jon Scott.
p. cm.
Includes bibliographical references and index.
ISBN 978–0–19–921983–4
1. Life sciences—Study and teaching.   2. Study skills.   3. Technical writing.
I. Scott, John.   II. Title.
QH315.J64 2009
570.71—dc22
2008051911

Typeset by Graphicraft Limited, Hong Kong
Printed in Italy by
L.E.G.O. S.p.A

ISBN 978–0–19–921983–4

1 3 5 7 9 10 8 6 4 2

To Ted, for keeping us both fit and healthy and
not giving us too many press-ups!

Preface

The transition from school or college to university is a very significant life event: for all students it marks the beginning of the next stage in their educational career and, for many, it may be their first experience of living away from home for any length of time. This transition, therefore, is associated with a whole range of new experiences in social and academic terms. Among these is a marked change in learning and teaching styles, with an expectation of increasing independence as a learner and development of a more mature, critical approach to the subject. Meeting the challenges posed by these changes requires the development of new skills in many areas of life, not least in the areas of study and communication. If you can develop effective study and communication skills early in your university career, you will facilitate your overall learning and help maximize your academic performance.

This book arose from nearly 10 years of experience in developing and delivering a first-year module in Study and Communication Skills for the students in the School of Biological Sciences at the University of Leicester. The structure of the book and its format are based on our experiences of delivering this course, in terms of recognizing where students experience difficulty in developing these skills and in learning from those students better ways of teaching and providing support for their learning. We hope that the future generations of students will be able to benefit from this guidance and we also look forward to learning more about the ways in which skills can best be developed.

S.J. and J.S.
University of Leicester

Acknowledgements

We would like to thank the students in the School of Biological Sciences at the University of Leicester who have helped to shape our work and, over the years, have experienced the evolution of the first-year module in Study and Communication Skills. We are also grateful to Maria Graal for helpful comments and support throughout the writing, and to Richard Clark who contributed significantly to the initial development of the module.

Finally, we thank Jonathan Crowe and Oxford University Press for their support throughout the project and for not harassing us too much when our timings slipped.

Contents

Chapter 1

Why are study and communication skills important?

⮑ Introduction

As the title suggests, this book explores the key study and communication skills that are needed by bioscience students beyond school level. This means that if you're studying, or about to study, biosciences this book can help. As you may have already gleaned from the contents pages, this book addresses a range of important topics, including:

- making the most of lectures;
- working with different information sources;
- choosing the right writing style;
- writing essays;
- writing practical reports;
- working in tutorials and groups;
- avoiding plagiarism;
- preparing scientific presentations;
- delivering scientific presentations;
- creating academic posters;
- using feedback;
- getting the most out of revision;
- getting the most out of exams.

We have written this book to help you in all of these areas and you may wish to work through them one by one. Alternatively, however, it may be that you have a specific issue that you need help with and so want to skip directly to the relevant chapter. Either approach is fine: we have written the book so that it makes sense when read as a whole text or as individual chapters. It is a good idea to read this chapter first though, as it helps set the scene.

Good study and communication skills are vital if you are going to do well at university; they will help you to study efficiently and effectively and so perform to the best of your ability in your chosen area of the biosciences. However, good study and communication skills aren't only useful for studying for your course; the kinds of skills you need to develop to do well in your chosen field of study have many parallels with the kinds of skills that will be useful throughout your career. Being able to research a subject, construct an argument, write a report, present information, manage your time, and plan your own development are all skills that are highly sought after by most, if not all, employers. So we hope this book won't just help you to do better in your degree, but will also help you to do well beyond university.

Incidentally, in this chapter we mainly talk about studying for a degree at university. However, this book is equally useful if you're doing another kind of bioscience course, whether at university or college.

1.1 'I did study skills at school. Why do I still need to read this book?'

Clearly, if you have got as far as getting a place at university you must be good at studying. Also, you were probably taught about study skills at school or college, and you will have certainly developed your own study skills during your time there. So why do you need to read this book? The main reason is this: studying for a degree at university is different from studying at pre-university level. There are numerous ways in which it's different. The important ones are as follows:

- studying at university requires you to be more independent than you were at school. You will therefore be expected to undertake much more self-directed study and so will need to learn to **manage your time well**;

- studying at university requires you to be more critical about your subject than you were at school, and not just accept everything at face value without question. You will be expected to identify differing views on a topic and develop your own views too. So you will need to learn to **think critically**;

- studying at university requires you to be more self-reflective about your ability and performance than you were at school in order to adapt and develop your current skills to better suit an undergraduate level of study. So you will need to learn to **develop yourself**.

Studying at university is, for most, a more challenging experience than school. This book will help you to meet those challenges and assist you in making the transition to become an accomplished undergraduate student. This chapter will focus on the above differences to highlight what we consider to be the foundational (or fundamental) skills you need to develop (and keep in mind) throughout your studies.

1.2 The contextual nature of skills: making the transition to undergraduate study

Skills are contextual. By this we mean that skills are learnt in a particular context, and the way you apply them is specific to that context. For example, when you learn to drive a car you learn in a certain way—usually with a driving instructor sitting next to you, in a car that's familiar to you, and receiving instruction on what to do. But just because someone has learnt to drive, it doesn't mean the person is capable of driving well in any and every situation. The chances are, initially at least, that the driving skills will be limited to a similar context to that within which they were originally learnt. Read through the following scenario to see what we mean.

- Imagine a friend of yours has spent the last few months learning to drive and they have just passed their test, passing with only two minor faults.
- If they offered you a lift, in a similar car to the one they had learnt in, would you feel safe catching a lift with them?
- If they offered you a lift, in a similar car to the one they had learnt in, but it was night-time, would you feel safe catching a lift with them?
- If they offered you a lift, in a similar car to the one they had learnt in, but they were going to drive on a busy motorway, would you feel safe catching a lift with them?
- If they offered you a lift, but they were driving an articulated lorry (!), would you feel safe catching a lift with them?

The chances are you would probably be more reluctant to accept a lift the more unusual the context became. This is because the more unfamiliar a context is to someone the more difficult it is to perform well in that context.

There are clear parallels between this example and making the transition to under-graduate study, which we are sure you will have spotted. You already have study and communication skills, probably ones that have served you well so far, but, as we have seen, studying for a degree at university is different from studying at pre-university level. Your skills will therefore need adapting and developing to fit the new situation you are encountering. The good news is that skills can be transferred from one context to another, so the skills that you have developed so far are by no means wasted. In fact, they provide a very useful starting point.

1.3 **Foundational skills**

As we mentioned above, the important differences between studying at pre-university and at undergraduate level reveal three foundational skills that are vital for students who want to perform well during their degree: managing your time, thinking critically, and developing yourself. These aren't the only skills—and subsequent chapters highlight many others—but they do underpin the ones that follow and so are worth drawing your attention to at this early stage.

1.3.1 **Manage your time well**

Many students, on nearing the end of their degree course, wish that they could have their time again. This is often because they feel that, given another go, they could do better. The reasons for thinking they could do better may be quite diverse, but common to many of them will be the notion that if they had managed their time more effectively they would have been able to perform more effectively, and get a better degree. Therefore, managing your time well is crucial to performing well at university. There will be many demands on your time—social as well as academic—and you will have more autonomy about how you choose to spend your time compared with school or college. It's important, therefore, that you learn to allocate your time appropriately. This is one of those things that is easier to say than actually to do; in theory it's straightforward, in practice it's difficult.

We have identified the need to manage your time well as a foundational skill because it is necessary for performing well in a wide range of academic tasks, but that doesn't mean we are going to deal with it in an abstract way—with a chapter just about time management, for example. Instead we address it in what we think is a much more helpful way: giving you advice about how to manage your time wherever it is relevant, in context—for example, when writing an essay, giving a presentation, or preparing for exams. As such, effective time management is a theme that occurs throughout the book, so look out for these sections in the chapters that follow.

1.3.2 **Think critically**

The second foundational skill that we want to draw your attention to in this opening chapter is the need to learn to think critically. Critical thinking is common to all academic endeavours and so it is important to understand what it is. Many people think of 'critical' as a negative term, and perhaps words like 'unfavourable', 'fault-finding', or even 'unkind' spring to mind. Being critical is, however, much more than this. If you have ever read a review of a film, book, play, or concert, the review will have been written by a 'critic': someone whose professional job is to be critical in its full sense. For example, the critique of the performance of a play comments on all sorts of aspects of that play, including the quality of the performances of the individual actors, the way the stage was designed, the lighting, and so on. Thus, the critic may have written in glowing terms about the leading lady but felt that the supporting cast was not up to the job. In this context being critical may mean being very positive and negative simultaneously, about different aspects of the same thing. However, it is important to note that this kind of criticism is subjective: one critic's view of the play may be very different from another's.

In the academic sense, the word 'critical' also doesn't have specifically negative connotations. In fact, in an academic context, being critical is a good thing and is something to be encouraged. However, there is one key difference between thinking critically in scientific terms and in terms of reviewing a play: in science, the critical thinking should be undertaken *objectively*. For example, if you are comparing two theories, you should compare the ideas being presented and evaluate the weight of evidence supporting them, using this evaluation to decide which is most tenable. You should approach this evaluation from a neutral (impartial) viewpoint: not looking for something to be right or wrong because you *think* it should be, but because the available evidence indicates it to be so. Critical thinking is therefore a process that requires the evaluation of evidence to come to a conclusion that is supported by the outcome of experiments or observations: it is an intellectual exercise that is fundamental to the way in which science works and develops: the basis of scientific method.

As with managing your time, this is not a book primarily about thinking critically. Rather we will address what it means to think critically when undertaking particular academic tasks—for example, being involved in lectures or tutorials, or when researching information for an essay or presentation. Look out for these sections in the chapters that follow.

1.3.3 **Develop yourself**

The third and final foundational skill that we have identified is the need to learn to develop yourself. This won't be an unusual concept to you as doubtless you will have been encouraged to engage in some kind of personal development planning at school. To make the transition from pre-university to undergraduate level, however, you will need to continue to develop yourself. Most universities will have a personal development planning scheme which you will be encouraged to take part in. Such schemes are designed to help you in two ways: to improve your academic performance, and to help you make plans for what you will do once you have graduated. It is a process of continuous improvement: thinking about and reflecting on what you have done in the past, and learning from this experience to influence in a positive way what you do in the future (asking yourself questions like 'what worked well?', 'what didn't go as planned?', 'how can I make sure that the things that didn't work well work better next time?').

Regardless of the format of the system you use, the important thing is that you reflect on your progress (to identify where you are doing well and where you need to improve) and make plans for your future development. You will have numerous opportunities for reflection, including:

- feedback on coursework;
- feedback on exams;
- conversations with your lecturers;
- conversations with your personal tutor;
- conversations with friends;
- time by yourself thinking.

It is important that you use these opportunities to assess how you are doing and then (just as importantly) decide what you are going to do about it. If you have done well on something, then celebrate it in some way (you probably don't need much encouragement to do this!) but also make sure you are aware of why you got a good mark so you can take that approach again. Equally, when you identify areas for improvement, plan what you need to do to improve. Reading this book is a good start.

✳ Chapter summary

Good study and communication skills are vital if you are going to do well at university. Undergraduate study is different from pre-university study in a number of important ways. The skills you have already gained up to this point will be useful, but you will need to develop and adapt them for the new context you now find yourself in. Managing your time well, thinking critically, and being committed to self-development underpin the other skills that we will now explore in the subsequent chapters.

Chapter 2

Making the most of lectures

⊃ Introduction

Lectures are a prominent and important feature of all undergraduate bioscience courses, therefore being able to make the most of them is an important skill. The lectures you have as part of your course will be many and varied; your lecturers will have different styles and different levels of ability in lecturing, and the lecture content itself will be of more or less interest to you and more or less demanding. Regardless of these variables, however, you need to make the most of lectures. This chapter will help you understand how to make the most of lectures by firstly addressing the purpose of lectures and then giving advice on how to:

- prepare before lectures;
- listen and make notes during lectures;
- follow-up after lectures.

2.1 The purpose of lectures

Lectures can be a very valuable resource as they can synthesize the views of several researchers and textbooks or provide new, and even unpublished, information. Lectures are often liked by academics as a means of communication because they are a very efficient means (in theory at least) of transferring lots of information to large numbers of students all at once. How efficient a lecture actually is at communicating information will depend on at least two variables:

- how well the lecturer prepares and delivers the lecture;
- how well the audience is listening and making notes.

Clearly, you can't do much about the first of these variables (how well the lecturer prepares and delivers the lecture), at least not in the short term. However, do make sure you feed back to staff your thoughts on how you found a particular module, usually via the end-of-module survey (if you think your feedback needs acting on more urgently, contact the module convenor directly). The second variable (listening and making notes) is clearly an area you can do something about; in fact, no one else can do it for you!

2.1.1 More than information transfer

Lectures are not merely about transferring information from the lecturer to the audience. In fact, we would argue that they are not even primarily about transferring information from the lecturer to the audience. If the purpose of a lecture was only to transfer information from the lecturer's brain to the students' notes, then lectures really have very little point because there are many more efficient ways of doing this; for instance preparing a handout, or suggesting particular chapters of a core text to read.

Lectures are of value because they add something in the communication of information that couldn't be achieved as well by other means. You will know this if you have ever tried to read a handout or a friend's notes from a lecture that you didn't attend; trying to understand the information out of context can often be very difficult. So if lectures aren't primarily about transferring information from the lecturer to the audience, what should they be about? The transfer of information is important; however, it is what you do with the information as you receive it, and afterwards, that really matters. This is where you, as a member of the audience, come in. It is perfectly possible, even sitting in a lecture that has been well prepared and is being well delivered, to disengage your brain and relegate yourself to being a mere passive recipient of the information.

As you will see later in the chapter, it is possible to not think about a lecture even while you are taking notes on it. The amount of notes you take isn't the measure of whether you have been paying attention; in fact, you might have a lot of notes because you haven't been paying attention! Doing something useful with the information as you receive it depends on processing the information effectively. This processing can't all take place within the lecture itself (even if you do listen actively and make notes appropriately), which is why this chapter covers preparation before lectures and follow-up after lectures, as well as what to do during lectures. Before we address these issues, it will be helpful to consider how lectures are different from what you may have experienced at school or college, prior to your undergraduate degree.

2.1.2 Differences between lectures and lessons

Lectures at undergraduate level will be different from lessons you had at school or college in a number of important ways. At the risk of generalizing, these differences probably include the following.

Class size

In years 12 and 13 at school you may well have been in a class of less than 30 students; at undergraduate level it's not unusual, especially in the first year, for class sizes to be up to 200–300 students (although typically, as you move through your course to the final year, the class sizes will get significantly smaller).

Anonymity

A bigger class makes individuals within the class more anonymous; it is easier in a larger class just to sit quietly and not really engage with the lecture content because you feel like no one will notice.

Relationship

Bigger class sizes mean that you are more remote from the lecturer, not only in terms of physical space but also in terms of relationship. The chances are that, in your first year of study, you probably won't get to know the lecturer very well and they probably won't even know your name.

Completeness

In school your lessons probably covered everything you needed about your syllabus; at undergraduate level, lectures don't tell you everything you need to know and so need to be supplemented with additional reading (more on this one later).

2.2 **Prepare before lectures**

You are already on a busy course with a lot of timetabled teaching hours—made up of lectures, tutorials, and practical classes—is preparing for lectures really necessary? We think it is. Before you discount this as something only for extremely keen and diligent students, bear this in mind: preparing for lectures doesn't have to take long and can actually save you a lot of time. It doesn't have to take long because you are not trying to cover all the material the lecture is going to cover (that's what the lecture is for), instead you are just giving yourself a framework to help you understand it better. It can save you a lot of time because if you have a framework to help you understand the lecture, you can process the information more quickly, thus making the listening easier, the note-making more selective, and the follow-up more focused. In order to know how to prepare for lectures, though, you first need to know some basic information about your course.

2.2.1 **Know your course**

In this context, by 'know your course', we mean finding out how your lectures relate to the course as a whole. This will involve finding out answers to questions such as:

- Do the practicals prepare you for the lectures or do they follow-up on material covered in the lectures?
- Are there opportunities to discuss the lectures in your tutorials?
- Is there any recommended pre-reading, if so, what is it?

Answers to these questions should be available in your module handbook, or altern-atively from your tutor. Additionally you will need to know the title of each lecture, or group of lectures, within any given module; this too should be in the module handbook which may also contain further details about what each lecture will cover.

2.2.2 Reading before the lecture

We have already noted that preparing for lectures doesn't have to take long because you are not trying to cover all the material the lecture is going to cover, but are just giving yourself a framework to help you understand it better. A framework will help you understand the lecture because it gives you somewhere to place information, and so relate it to other information, rather than trying to understand it in isolation (which is always very difficult).

The framework can be quick to create because you only need to understand basic structure and terminology, rather than detail. A good way of achieving this is by looking up the relevant chapter in the core text for the module (this is where you need to know the title of each lecture) and scanning through it to familiarize yourself with the main themes. This doesn't require you to read the whole chapter, or read it in detail, just scan through all or some of the following:

- the chapter overview (if there is one);
- the chapter headings and subheadings;
- the introduction to the chapter;
- the conclusion or summary to the chapter;
- figures, diagrams, or tables (which often summarize a lot of information succinctly);
- also, look out for terminology that you are unfamiliar with and try and find out what it means.

Core texts are not the only source of information for pre-reading, however; your own notes will be useful too. Assuming you are not on the first lecture of a module, your notes from previous lectures in the module will be an important source of information. Again, you don't have to read all the notes in detail, just scan through them to remind yourself of what was covered previously, but, when you come to a section that you don't under-stand, spend more time on it and check up on the topic in a textbook or other reference source, or ask your friends.

If you have prepared for a lecture it makes listening and note-making much easier; which are the subjects of our next two sections.

2.3 **Listen actively during lectures**

This section is deliberately titled 'listen actively' because there is a difference between just listening and listening actively. Listening actively suggests that you are alert, attentive, and ready to engage with the content of the lecture, as opposed to just being there. Clearly, the quality of the lecture (how well it is prepared and delivered) will have an effect on how easy it is for you to listen, but listening actively means that you are willing to work hard at listening regardless of the quality of the lecture. Additionally, we have already identified that preparing for a lecture makes listening actively much easier, because you have a framework within which to place new information and you will at least recognize the terminology used.

2.3.1 **Identify your priority**

There will be some lectures when it will be difficult both to listen and to make notes, either because of the complexity or newness of information being communicated, or simply because of the amount. In such situations you will need to identify what your priority is; is it to listen or is it to make notes? Deciding which your priority is will depend on whether you are more concerned with trying to **understand the information** or **collect information**. If your priority is understanding the information then you will need to:

* focus on listening;
* make only brief keyword notes;
* follow-up the lecture by making detailed notes.

If, however, your priority is collecting information, then you will need to:

* focus on making notes;
* make more detailed notes;
* follow-up the lecture by reviewing your understanding of the content.

When making a decision about your priority, also consider what resources are available to you after the lecture. It is usually relatively easy to get hold of appropriate textbooks or perhaps borrow a friend's notes, and you may well also have access to lecture handouts; however, it is much more difficult actually to experience the lecture again. Even if you can listen to a recording it is rarely as good as actually being there. This would suggest that active listening, and therefore understanding, should be more important than making notes. Remember, however, you don't necessarily have to keep to the same strategy all the way through a given lecture; you could switch between the two depending on the material (more on this in section 2.4).

2.3.2 Listen for structure

Making notes is always easier if you have an awareness of the structure of the lecture. Being aware of structure enables you to be more selective, and therefore more focused, in your note-making because it gives you an indication of what the important bits are. That's not to say that the other bits are unimportant, it's just a recognition that in terms of understanding there are certain elements of a lecture that are helpful to grasp in order to understand the rest of it.

Sometimes the structure of a lecture is very clear, other times it is less clear. It is important, therefore, to know what the structural clues might be. The most obvious and probably the most common way of making the structure of a lecture clear is for the lecturer to outline what the structure is going to be at the beginning of the lecture, either as a list of headings or as a potted summary of the content. If a lecturer does outline the structure, make sure that you make a note of it straightaway (this applies equally whether your priority is focusing on understanding or collecting information), as it will give you a sense of direction and help you to anticipate points or take up the thread of information again should you get lost.

Additionally, during a lecture the lecturer will probably give you some cues, or 'verbal signposts', these include statements such as:

- 'I shall now discuss . . .'
- 'My next point is . . .'
- 'Finally . . .'
- 'In conclusion . . .'
- 'To summarize . . .'

These signposts identify a new point or stage in the lecture and you should show this in your notes accordingly. Other signposts include: pausing to indicate a new point or summarizing what has been said prior to moving on.

There are other, more subtle, verbal signposts which can help you structure your notes; you will need to listen for these. Examples include:

- 'On the other hand . . .'
- 'Others have argued . . .'
- 'Turning now to . . .'
- 'Alternatively . . .'

Other words and phrases indicate that an illustration is about to be given:

- 'An example of this is . . .'
- 'This can be seen when . . .'
- 'Evidence for this can be found in . . .'

Your ability to listen actively, and in particular to listen for structure, will improve with experience. As you improve you will be better able to spot digressions or additional examples and adjust your note-making accordingly, which brings us on to the next section.

2.4 Make notes appropriately during lectures

We commented earlier (section 2.1.1) that the amount of notes you take isn't the measure of whether you have been paying attention; in fact, you might have a lot of notes because you haven't been paying attention. We therefore need to draw a distinction between making notes and taking notes. We have deliberately referred to making notes so far, as opposed to taking notes, because we think there is an important difference. If you take something you are just the recipient of it, so in the case of lectures you just sit there and write down what the lecturer is saying (or worse, just walk away with the handout).

On the other hand, if you make something you are involved somehow in the creation of it and you bring something of yourself to it, so in the case of lectures you create notes that are unique to you and are produced by you engaging thoughtfully with the content of the lecture. Making notes, therefore, is an active activity whereas taking notes is usually a passive activity. As you can see, there's quite a difference. That's not to say that making notes won't involve you writing down word-for-word what a lecturer is saying on occasions; the important thing is that you make notes appropriately. As we have already seen in section 2.3.1, this will depend a lot on what your priority is; whether it's to understand the information (and so you focus on listening) or to collect information (and so you focus on note-making). In reality, although it is difficult to focus on both listening and making notes at the same time, it is not that you just do one or the other during any given lecture; rather the two go hand-in-hand. You will be constantly switching between listening and making notes all the way through a lecture, but how much you do of each should reflect your priority, as described in section 2.3.1 above.

2.4.1 Know why you are making notes

We have already encouraged you to think about what your priority might be when in lectures, and understanding information and collecting information are two key reasons for making notes. However, there are other reasons too. Making notes appropriately in lectures can also help you to:

- concentrate better;
- remember the content better (in the short-term at least);
- think about questions you want to ask;
- highlight areas of interest.

Note-making, therefore, isn't just about having something to refer to later; it is part of the learning process itself. This is another reason why borrowing a friend's notes is a poor substitute for actually being there and experiencing the lecture for yourself, because so much takes place in the lecture, and in your head, that even really good notes can't really capture the whole event.

2.4.2 Know how to make notes

The ability to make good notes is a skill that develops with practice, so don't expect to be an expert at it straight away. Clearly you will have made notes before at school or college, but making notes in lectures in an undergraduate setting is different for a number of reasons, as highlighted in section 2.1.2. So improving your note-making skills comes partly through practice, but there are some important principles to bear in mind too; these include:

- using structure in your notes;
- using your own words;
- using fewer words;
- using abbreviations;
- using space;
- using colour and image;
- using handouts;
- organizing your notes.

We will address each of these principles in turn.

Use structure in your notes

Notes that are lacking in structure will be much more difficult to understand, especially after the lecture, than notes that have a good structure. Imagine you were looking at some lecture notes a few days after a lecture that looked like the ones in Figure 2.1. Notes like these are difficult to understand partly because they are lacking in structure.

Alternatively, the notes represented in Figure 2.2 have better structure and so should be much easier to understand. The main purpose of structure is to make clear which information is important. It is crucial to note that identifying important information is much easier during the lecture because lecturers will use their tone of voice, pace, and many other devices to provide emphasis to the material, thus giving you an indication of what aspects of the information are most important. After the lecture, however, it is very difficult to recall this kind of detail.

To give your notes structure, you should:

- use headings to order information;
- give each point a new line;

FIGURE 2.1 Poorly structured lecture notes.

Respiration – Prof Smith   22 Feb

Inspiration – depends on contraction of diaphragm – flattens – and of external intercostals. Cause increase in thoracic vol., reduces pressure, draws air in.
Expiration – passive process – elastic recoil of lung tissue. Air flow is tidal, air in lung is not the same composition as air outside – has more CO_2, less O_2 'cos some remains in lung at end of expiration – even if forced. Inspiration has to do work – overcome elastic tension of lung tissue and surface tension of fluid lining lungs. ?role of surfactant??
Gas exchange – occurs in terminal bronchioles and alveoli. O_2 dissolves in fluid and diffuses across lung wall to blood vessels – down concentration gradient. O_2 poorly soluble so needs large surface area for dissolving. CO_2 is better, goes in opposite direction. Need short diffusion path.

FIGURE 2.2 Well-structured lecture notes.

Respiration – Prof Smith   22 Feb

Inspiration*
- Dia. + ext intercostals contract
  - thoracic vol ↑ + pressure ↓
  - air drawn in

* has to do work – overcome elastic tension of lung tissue and surface tension of fluid lining lungs. ?role of surfactant??

Expiration
- passive process – elastic recoil of lung tissue
- air flow is tidal
- air in lung is not the same composition as air outside – ↑ CO_2, ↓ O_2 (some remains in lung at end of exp.)

Gas exchange
- occurs in terminal bronchioles + alveoli
- O_2 dissolves in fluid and diffuses across lung wall to blood vessel – down concentration gradient
- O_2 poorly soluble so needs large surface area for dissolving
- CO_2 is better, goes in opposite direction. Need short diffusion path.

- highlight examples and illustrations in an appropriate fashion;
- use diagrams to summarize information;
- make clear when sections of your notes are digressions from the main points.

Use your own words

One of the reasons why it is important to try to use your own words, when making notes in lectures, is because it will help you (or perhaps force you) to understand the content of the lecture better. If you are trying to put information in your own words, then you must process it as you hear it in order to express it in your own words. It also helps you to make notes, as opposed merely to take notes (as explained above). You may have experienced occasions in lectures when you have been able to write down what the lecturer is saying without actually thinking about what is being said—this illustrates how passive the process of note-taking can actually be. How much you try to put information into your own words will depend to a certain extent on your priority; but whether your priority is to understand or collect information, putting that information in your own words will help you understand it better. This doesn't mean that you have to put absolutely everything in your own words; for example, there are two particular occasions when it is important to record the precise wording:

- when you are recording a quotation that the lecturer is referring to—in which case make this clear in your notes by using quotation marks;
- when you don't understand what the lecturer's words mean—in which case make this clear in your notes by adding a question mark in the margin (for example) as a reminder to follow-up the point later.

Know how much to write down

There are two potential problems related to how much information you write down; writing down too much information or not writing down enough. How much you write down will again depend on your priority (understanding information or collecting information), but it is also influenced by your own level of confidence. Under-confidence tends to lead to writing down too much, whereas over-confidence tends to lead to not writing down enough; neither is ideal. Among first-year undergraduates probably the more common tendency is to write down too much information, but, whichever your tendency, the following suggestions will help:

- remember that lots of notes don't necessarily equal good notes;
- look and listen for the important points—these are often the structural parts;
- use keywords to represent points or ideas concisely;
- add brief details of any examples or evidence that support a point.

Use abbreviations

Using abbreviations can be a real time-saver (as you will know from text messaging). Use standard abbreviations, subject-specific abbreviations, and your own abbreviations for common words. The important thing is to be consistent and to ensure that your notes are still comprehensible; don't use so many abbreviations that your notes turn into a short-hand transcript—these can be very difficult to decipher when your memory of the lecture has faded.

Use space

It can be tempting, in an effort to save paper, to try to cram as much information onto a page as possible, but this will create difficulties for you both during and after the lecture. Notes that are densely packed are difficult to review and difficult to make additions to at a later stage. It is therefore helpful to use space in your lecture notes to make them easier to review and easier to supplement with additional material. To create space make sure you put each point on a new line (this also helps represent structure) and leave gaps for additions or corrections, especially if you think you may have missed or don't understand something.

Use colour and image

We have already identified in the section on *Use structure in your notes* the value of highlighting important points by using structure, but this can be further enhanced by using colour and image too, as shown in Figure 2.3. It can be useful to highlight in colour key points and to use images or diagrams as a quick way of summarizing a concept or idea. Sometimes such images or diagrams will be used by the lecturer, in which case, if it is useful, copy it down (although see the points below in *Use slides and handouts effectively*). On other occasions you will think of ways to represent visually something that is only communicated verbally; this can also be a very helpful thing to do, but make sure you record sufficient information to be able to understand the concept or idea at a later stage.

Use slides and handouts effectively

Slides and handouts are a very valuable source of information. Not all lectures will have handouts but most will have slides. Assuming that copies of the slides are available to you (either before or after the lecture) then the principles of using them effectively are similar.

- Put your name on the handout—everyone's will be identical (initially at least), and you will want to know which one is yours.

- Put a date and lecture title on the handout (if there isn't one on it already) so you know which set of lecture notes it belongs to later.

- When making notes, think about what is contained in the handouts or slides and don't copy down in your notes things that are already in the handouts. This will enable you to listen more actively and so focus on understanding.

- Make notes on the handouts themselves—highlight important points, add comments, write down any questions you have.

FIGURE 2.3 Using colour and image in notes.

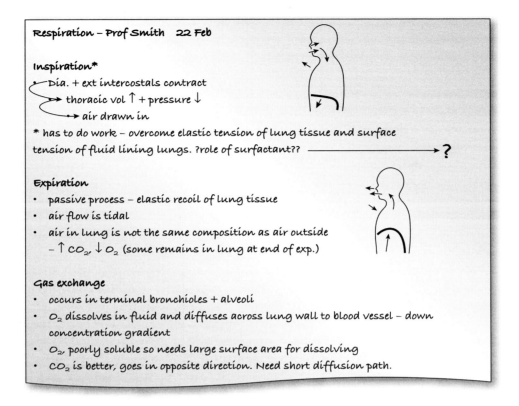

Respiration – Prof Smith   22 Feb

Inspiration*
• Dia. + ext intercostals contract
  →→ thoracic vol ↑ + pressure ↓
  → air drawn in
* has to do work – overcome elastic tension of lung tissue and surface
tension of fluid lining lungs. ?role of surfactant?? —————————→ ?

Expiration
• passive process – elastic recoil of lung tissue
• air flow is tidal
• air in lung is not the same composition as air outside
  – ↑ CO_2, ↓ O_2 (some remains in lung at end of exp.)

Gas exchange
• occurs in terminal bronchioles + alveoli
• O_2 dissolves in fluid and diffuses across lung wall to blood vessel – down
  concentration gradient
• O_2, poorly soluble so needs large surface area for dissolving
• CO_2 is better, goes in opposite direction. Need short diffusion path.

- Don't fall into the trap of thinking that because you've got a handout or slides that you don't need to listen much—listening actively during the lecture and making appropriate notes (on your handouts and in your own notes) will save you a lot of time later and also make it easier to understand the material.

- File your handouts in an organized fashion (which brings us neatly onto the penultimate point in this section).

Organize your notes

It is important that you organize your notes effectively, as this will make it much easier for you when you follow-up the lecture at a later stage or use the notes for background reading for an essay, or when revising for exams. The simplest, and in our experience the most effective, method is to use A4 paper for taking notes and to store your notes in a ring binder, along with the relevant handouts. To make ordering your notes easier, begin notes for each lecture on a new piece of paper and give them a clear heading of the lecture title, date and name of the lecturer. Also add page numbers so you can order the pages easily. To store your notes and handouts in a ring binder you will need to hole-punch them—do this before they get lost!

Compare your notes with other people's

Finally in this section, compare your notes with other people's. Remember, we said earlier that the ability to make good notes is a skill that develops with practice; it is also a skill that develops as you see how other people do it. Comparing notes can be a helpful exercise because it can help you to:

- identify and fill in any gaps that you might have in your notes;
- discuss the content of the lecture and clarify your understanding;
- identify how the lecture relates to the rest of the module.

2.4.3 Know the common problems and how to address them

We have addressed the ways to help you to make appropriate notes, but there will always be times when, for a variety of reasons, concentration dips and you begin to get left behind in the lecture. The trick here is not to despair and give up!

Failing concentration

You are much less likely to find your concentration straying when you use an active approach to note-making. By using your own words, and by using space, colour, and image, note-making will become a busy but interesting activity. If you do miss some points because your attention strays, then just leave a space in your notes and check it out with the lecturer or another student later.

Being left behind

You may find that information is being delivered too fast for you to write down. If points pass you by, then leave a space and compare your notes with another student's. Doing some background reading for the lecture will help you to keep up as the information won't be entirely unfamiliar to you. Sometimes you can get lost because you don't understand the material that is being delivered. This may be the case for the occasional point or even for a large section of the lecture. Rather than giving up on the lecture, write a series of questions that you can try to follow-up later.

2.5 Follow-up after lectures

At school or college, lessons usually cover everything a student needs to know about their syllabus, but at undergraduate level lectures don't tell you everything you need to know and so need to be supplemented with additional reading. It is vitally important that you are aware of this because when you are revising for exams, if you only revise the lecture material your knowledge of a subject will be inadequate. We will deal with this more in Chapter 13, *Getting the most out of revision*.

2.5.1 **Know what kind of follow-up is required**

In terms of the information covered, lectures are commonly used either to offer a broad overview of a subject (in which case your job is to fill it the detail) or to deliver detailed information on a subject (in which case your job is to fill in the background). There are therefore two issues to consider when identifying what sort of follow-up a lecture requires: does the subject require more depth (because the lecture gave an overview) or does the subject more breadth (because the lecture gave detail)?

2.5.2 **Ask questions**

Don't be afraid to ask a lecturer for clarification either during the lecture or afterwards. This can take a bit of courage, especially putting your hand up and asking a question in a room full of people, but it is very unlikely that you are the only person in the room think-ing that particular question. Also, lecturers will often welcome such questions as they provide a bit of stimulus for them as well as indicating that people in the room are paying attention. Additionally, use tutorials to clarify or discuss material from the lectures.

2.5.3 **Review your notes**

Review your notes as soon as possible after a lecture, and make the most of your review by:

- highlighting points which seem particularly important or central;
- adding any details which you can remember from the lecture;
- showing links between points;
- correcting any mistakes;
- adding questions to highlight areas you don't understand or need further informa-tion on.

✳ **Chapter summary**

Making the most of lectures is an important skill for undergraduates. There is much you can do to improve your ability to listen and make notes; this includes things you can do before, during and after lectures. The things you can do include:

- be prepared: get to know the course structure and do pre-lecture reading;
- be organized: have a system for storing notes and take a selection of pens and pencils to the lecture;
- listen for structure: watch out for signposts that help you follow the lecture;

- be brief: try using key words and phrases as much as possible so the emphasis is on listening and understanding;

- let note-making be an active process: summarize in your own words, make good use of space, colour, symbols, and images;

- leave gaps: if you miss a point, don't get further behind by panicking about it, just leave a space and check it out with another student after the lecture;

- actively review your notes after the lecture, making additions and corrections as you read.

Chapter 3

Working with different information sources

➔ Introduction

In previous generations, students were said to go to university 'to read for a degree', the implication being that most of their working time as students would be spent reading up on the subject they were studying. Even in subjects like the biosciences, that have a strong emphasis on practical work, reading is still a very important, if often underrated, part of studying.

What is the purpose of reading? This is an important question because there are several reasons why you will need to read during the course of your degree, and the approach that you take will depend on what your goal for the reading is. Some key reasons for reading include the following:

- background reading to find out more about the subject in general;
- following up on lectures, or other teaching activities, in order to help you understand the subject matter you have been taught;
- following up on lectures to develop a deeper and broader appreciation of the subject matter;
- researching topics as part of the preparation for a coursework assignment such as writing an essay or delivering a presentation.

These activities require different approaches to reading because there are very different goals associated with them. For example, the sources that you use and the strategies employed for finding the information will differ, depending both on the goal for your activity and the level you have reached in your course.

In this chapter, we will explore how to get the most from texts, journals, and the internet, and how to evaluate the academic value of a variety of sources. To do this, we will review:

- the types of publication available and how you might use them;
- using internet sources;
- referencing and bibliographies;
- different reading goals and how you approach them;
- note-making strategies.

3.1 Different types of printed publication

There is a wide range of types of publication available for use at different levels of study. Each of these has different characteristics and will probably be used in different ways. In this section we will consider the main forms of publication, some of their key characteristics as study aids, and suggestions as to their use. Further consideration of the use of publications and internet sources are given in the sections on note-making and reading goals.

A fairly simple way of categorizing the types of publication you might want to use for your studies is as follows:

- textbook;
- specialist topic book/monograph;
- general science journal;
- specialist review article;
- research paper.

Let's consider each of these in turn.

3.1.1 Textbooks

The textbook is probably the form of publication with which you are most familiar from your studies at school or college. Textbooks still form a very important resource for students at university, predominantly in the first and second years of study. Whereas many of your textbooks will have been broad-based (for example, a textbook covering all of biology), the books you will be using, even in your first year of study, will be more specialized, focusing on a specific sub-topic of biology such as biochemistry, genetics, or physiology. The principle is the same, however: the textbooks offer a broad-based foundation to the study of a subject.

Buying textbooks

Even within the individual topics such as biochemistry or physiology, if you go to a bookshop or search online you will find a bewildering array of books available to buy. It is important to be cautious when purchasing books otherwise you can find that you have spent a lot of money very quickly, sometimes on buying books that will not be of great use to you. You have clearly taken the first step in the right direction by buying this book, now you can proceed with caution!

- Most universities publish lists of recommended books for specific modules or courses, so start off by consulting the lists for the modules you are taking.
- Be selective: for example, is it worth buying a book for £40 or £50 for a subject area that you have to take as a single module in your first year, but which you do not intend to continue with later on in your course? If the university library has a number of copies available, you can probably manage without buying one for yourself.

- Having decided you do want to buy a book, you have to choose which one. It may be that your reading list only suggests a single core text, which makes life easy. But there may be a choice of recommended texts, perhaps covering the material to different levels and at fairly different prices. Again, the library is a useful starting point. Have a look at the different texts in the library: you will probably find that they are different in terms of their layout, the way diagrams are used, the level of detail, etc., so you may find it helpful to test-read parts of the book to find which style suits you best. You may also find that they differ in terms of the additional features they offer: for example, is there an associated website with more information or access to self-evaluation tests? Is there an accompanying CD or DVD?

- Wander past the student notice boards of your department or students' union and you will probably see numerous adverts for second-hand books. There may well be a second-hand bookshop. The books should be cheaper than buying brand new so you could save money, but you need to be cautious. For example, check that it is the current edition of the book that is on sale. If there is an accompanying CD, is it still there? If the book has an associated website, make sure that you can still access it: many sites have pin number access, which is unique to a single user and cannot be transferred if the book is sold on.

Using textbooks

Textbooks commonly provide a broad overview of a topic and can have several uses as study tools. Think about the reading goals that you have and that will guide you in the way to use your textbook. Some possible examples could be:

Goal 1

I need to broaden my knowledge and understanding of a specific topic, for example the function of the kidney.

Action 1

In a general physiology textbook read through the chapter(s) on the kidney, checking on any learning outcomes identified in the text. Make abbreviated notes to aid your learning (see section 3.3 on note-making).

Goal 2

I have just had a lecture on the physiology of the loop of Henlé and I don't understand it!

Action 2

Read through your lecture notes and any accompanying handouts. Look up the loop of Henlé in the index of your physiology textbook and read through the relevant section slowly. Then read it again, making notes about the organization of the loop and the way it functions. It might help to go to the library and read about it in a couple of different textbooks. You should also ask the lecturer for help, but it is a good idea to try to think it through for yourself first.

Goal 3

I have just been given an essay to write on the regulation of ion balance by the kidney. Where do I begin?

Action 3

Read through any relevant lecture notes and handouts first. Then read the sections in your physiology textbook that deal with ion balance, making detailed notes about each of the key ions in turn (e.g. Na^+, K^+, Cl^-, etc.). Make sure you put the notes in your own words and include the source reference (see section 3.4). Having got the general picture, you may then need to refer to more detailed sources such as a specialist topic book, review articles, or research papers.

As with any source of information, textbooks do have limitations, as follows.

- Because most textbooks offer a fairly broad coverage of the subject (e.g. physiology, or ecology, etc.) they may well only provide a fairly superficial overview of a topic. This is important because it may be just what you need for giving you the background to the topic, or for helping you understand lecture materials in your first and second years. They almost certainly will not, however, provide enough detail for the later parts of your course, or for researching material for coursework.

- Textbooks are often written by a small team of authors, who will not be specialists in all of the topics covered in the textbook. As a result, the quality of the coverage of the different topics may vary. For some texts, the book is prepared by editors, with subject specialists writing the individual sections. These are more likely to have a more even quality of coverage of the topics.

- Textbooks do go out of date, and this is likely to happen faster than for your school or college textbooks. It can take several years for a textbook to be written and published, and then there may be a few years more before the book is revised and a further edition published. Although this may not cause problems when used for basic reading (since the key principles of a topic are unlikely to change significantly), if you require more specialist information, then it is more important that the book you are using is up to date. So, when you are scanning the shelves of your library for textbooks, make sure that the books you select are not more than 2 or 3 years old.

3.1.2 Specialist topic books/monographs

As their name suggests, these are books that focus on a specific topic within the subject. These usually take three forms:

- An advanced textbook dealing specifically with a fairly narrow topic, such as kidney function. This will clearly be more detailed in approach than the broader textbook on physiology but will be written in a similar style. It will almost certainly have been written by a specialist, or group of specialists, within the field.

- A monograph usually takes the form of a long article or a short book on a specialist topic. This will typically be narrower in focus and at a more specialist level than the advanced textbook.

- At the research level, books are quite often based on the papers presented at a specifically themed conference. As with journal research papers, the individual papers may be specialized, being written for a readership of research scientists within the field. However, the grouping together of a whole series of papers on a specific theme may well provide you with a valuable starting point for your research for an essay topic. These books also frequently contain useful overviews of groups of papers in review form.

As with using textbooks, you need to make sure that the information you are looking at is up to date; indeed, the more specialized the book or research paper, the faster it is likely to go out of date. All of these, however, make valuable resources for advanced reading around a topic to increase your depth of knowledge, and they can also make very useful starting points in research for essays or similar assignments.

3.1.3 General science journals

There are several general science journals, such as *New Scientist* and *Scientific American*, that publish review-type articles aimed at the general interest reader with a scientific background. These types of journal can be very useful in providing overviews of a specific topic and also for keeping you up to date with wider developments in science as a whole.

3.1.4 Specialist review articles

Many research journals include review articles as well as the papers describing the results of novel research. There are also numerous journals dedicated to reviews in the biosciences, such as the '*Annual Reviews in . . .*' series or the '*Trends in . . .*' series.

Recent review articles are a gold mine for students seeking information about a specific topic. These articles are typically written by a well-known specialist in the field who brings together, in a single article, a synthesis of a large number of individual research papers. The approach of these articles is usually designed to enable someone with some knowledge of the field, but not necessarily at a very detailed level, to derive a picture of the current ideas and developments within that field without having to read all the individual research papers.

Review articles are very useful for providing more detailed information about a topic, for example for supplementing lecture material, particularly at second and third year level. They also often make very useful starting points in research for coursework assignments, both because of the overview they provide and because of the references made to recent research work in the field. Your lecturers will almost certainly include some review articles in the reading lists for your courses, particularly in the later years of your degree programme.

3.1.5 **Research papers**

Research papers published in academic journals are the dynamic bedrock on which current knowledge is based. These papers are written by scientists describing their latest discoveries or developments within their field of research. In any honours degree programme, there will be an expectation that you will:

- read recent research papers in order to supplement the lecture material;

- use them in order to prepare your coursework assignments;

- use the information you have acquired from reading research papers to support the arguments you present in examination essays (this will be particularly true for finals examinations).

As with the review articles, your lecturers will certainly include recommended research papers (some of which they have probably written themselves!) in the reading lists they give you to supplement the material covered in the lectures. The fact that someone who is an expert in the field has recommended a given paper suggests that the paper has merit as a piece of scientific writing and that the findings are likely to be reliable. In the same context, papers for publication in scientific journals are subjected to a peer review process: in this process, reviewers (usually two), who are also acknowledged experts in their field, are asked to read the paper and to comment on its scientific value, to rate its importance in the field, and suggest any amendments they think should be made. Although this is clearly not a foolproof process, and is certainly not an excuse for you to stop thinking critically when reading, it does give these papers a stamp of authority and reliability.

Research papers are the most specialized of all the standard sources of information and can therefore be quite intimidating and hard to tackle: for example, they will probably use terminologies and be based on ideas which you do not understand very well. Research papers are usually in a highly stylized format, presenting information in a way which you are probably not accustomed to. Finally, it must also be admitted that they don't always make the most interesting bedtime reading! So, if you are going to invest time and effort on reading research papers, you need to be very clear about your reading goals so that you get the most benefit from your input.

The structure of research papers

Most research papers are written in a standard format, comprising seven main elements (see also Chapter 6).

- *Title*. The title is a short descriptor of the paper; as such it will give you a good idea of whether the subject matter is relevant to your needs.

- *Abstract*. The abstract is usually about 200 words in length and provides a short summary of the paper. In particular this will include the aims of the research, the key results, and a brief commentary about the conclusions. It may be that this is all you need to read in order to glean the key points, for example to support the argument in an essay.

- *Introduction.* The introduction is the section of the paper where the authors set out the background to the study. This is often very useful both for helping you to understand the basis of the topic and the reason why the researchers undertook the investigation they are describing. As such it will normally include a useful summary of the preceding research and an analysis of the questions still to be answered. It usually culminates in setting out the aims of the study.

- *Methods.* The methods section describes the way in which the research was undertaken, including details of the experimental protocols and the procedures for analysing the data. Unlike the type of report that you might write (see Chapter 6) the protocols are often not described in full but rather refer to previous papers in which they have been described, for example:

> The muscle preparation was stained according to the method of Barker and Ip (1963) . . .

This form of summary information can be fine for people working in the field who already have a good knowledge of the literature. For the student trying to understand the paper, however, this can be a nuisance because it means further searching out of papers in order to find the actual description of the method, particularly if, as in this case, the method is an old but standard technique.

- *Results.* In the results section the authors will describe the results that they have obtained from their research and will also display them in different formats, for example as graphs, tables, or other illustrations. Where quantitative data are being presented this will also be analysed using statistical techniques. Note that this section is descriptive and does not normally include any discussion of the results or reference to the findings of other researchers.

- *Discussion.* This section, as its name suggests, is where the authors interpret the findings of their research both in terms of the specific experiments they have undertaken and also in the context of the current hypotheses in the field, as developed by other researchers. This section should link back to the Introduction where the aims of the project were set out. Again, this section makes very useful reading as it links what has been reported in the paper to the existing research literature.

- *References.* This comprises a list of all the research papers, etc., that were referred to in the text of the paper. The reference list is often very useful in providing a source of further reading to help you fill in the background of your topic.

Reading a research paper

From the reading list supplied by your lecturer and your own searching of databases, you will probably have collected a frighteningly long list of research articles on a topic. Unfortunately (and contrary to some popular opinion) the acts of downloading a paper or getting a copy from the library do not also transfer the contents of the paper to your brain.

You actually have to read the paper in order to extract the information from it. You also know that there is no way in which you have enough time to read all these papers.

So what should you do? This is where you need to be selective in your searching (see section 3.2) and think about your reading goals to decide what you really want to get from your reading. For example, is it some specific, factual information to add in to an essay? Or is it a detailed understanding of the topic as the background to a final year research project? It may be that just reading the abstract will suffice, or you might need to read the whole paper. These different approaches can be evaluated by careful consideration of your reading goals.

Goal 1

I am reading papers to find specific information for an essay or practical report.

Action 1

In this case, you are looking for specific information and do not necessarily need to read the whole paper. Check that the title appears relevant and then read the abstract: this should summarize the key points of the paper and may by itself be sufficient for what you need, particularly if the main use of the reference is to confirm a minor point in the essay. However, if the information you require is more detailed, then you will need to go deeper into the paper and use some of the strategies described below.

Goal 2

I am reading papers to support the lecture material on a module.

Action 2

For this type of reading, you may often begin from a reading list that you have been given along with your lectures. Therefore, you can start by selecting the papers from that list that appear directly relevant to the topic you are reading up on. As with Goal 1, you may be able to be quite selective in the extent of your reading, particularly if there is a specific point that you want to know more about. Alternatively, you may also wish to read the papers fully in order to appreciate the wider aspects of the topic. Try to make sure, before you start reading, that you are clear in your own mind why you are reading that paper and what you want to get from it.

Goal 3

I am preparing the critical literature review for my final year project.

Action 3

For this type of exercise you will probably need to read the paper carefully, especially if you are comparing the hypotheses put forward by different authors. As with all your reading, when you get to a section relevant to your reading goal you should make careful, referenced notes about what you have learnt from each paper. This is particularly true when engaged in a long-term piece of work such as your final year project because you may well need to be able to go back to the sources to check, or add to, what you have written.

So, how do you find this information if it is not taken from a reading list? The most common approach is to use internet-based searches, which we describe below.

3.2 Using the internet

Anyone who has looked on the internet will be aware that there is a vast, bewildering wealth of information available at the click of a mouse button. Some of that information is accurate, up to date, authoritative, and really useful. A lot of what you may unearth, however, is none of those things. So, while online searches can be very productive, you need to have a clear strategy for directing your search, for filtering the material that is available, and for being able to evaluate that information.

Let us say that you have been set an assignment on the mechanisms underlying muscle fatigue and you need to find out some recent information on the topic. If you need to start by finding out basic information, then it is probably best to start with lecture notes and handouts, progressing on to textbooks and reading lists as described above. When you have exhausted these, you need to undertake your own specific searches.

There are two main starting points: you can look specifically for articles in academic journals, or you can use a search engine to explore the whole of the web. In this exercise, we will explore both approaches for comparison.

3.2.1 Academic journals

When you are searching for specialist scientific information, then academic journals should be one of your main sources of information. Almost all universities and colleges subscribe to electronic journals and also to databases that search the academic literature. Through your own library's catalogue, you should be able to find what journals you will have direct access to via an online link. The library catalogue should also provide links for you to use databases for your search. Examples of such databases that are commonly employed by bioscientists are Web of Science, Ovid, EMBASE, PubMed, and Medline. Different databases cover different ranges of journals and so they vary in size and in terms of the number of journals included. Some, such as Web of Science, are relatively broad, others, such as Medline, are more specific, only including journals likely to be of relevance to medical research. You will need to select which database you use; it may be advisable to seek direction from a university librarian who will have a good knowledge of the databases your university subscribes to and the range of topics that they cover.

These databases allow very sophisticated searching strategies using a range of criteria. The most common strategy is likely to be based on a set of keywords that you think are core to the topic. If you have been given the name of a specific researcher, you could also search using the author's name. The advanced techniques of searching are beyond the scope of this book, and you should consult with an adviser from your library, but we will look at some simple strategies for searching, which will probably be adequate for much of what you want to do.

You need to find out about the mechanisms of muscle fatigue so you could start with a crude search in Web of Science and simply enter the words 'muscle fatigue'. This comes up with more than 8000 research papers. Similarly, if the same key words are entered into EMBASE, more than 5000 articles are listed. In both cases, you probably don't have time to scan the titles, let alone read the articles! So how can you refine the search strategy to pull out the articles that will be of value? There are several basic options here:

- choose specific types of article;
- select only more recent publications;
- refine your search terms.

See Figure 3.1 for an example.

FIGURE 3.1 Example of page 1 of the results from an EMBASE search, searching on the term: muscle fatigue.

Choose the type of article

Most search engines allow you to specify what type of article you would like to select. You can do this either by specifying what types you want to *select* or what types you want to *exclude*. For example, you can specify that you only want review articles, or you can exclude meeting abstracts (because they are likely to be short and very narrow in focus). If we only specify review articles, then the searches come up with about 200 articles, but not all of these will be directly relevant because of the breadth of the search terms used (see below). Furthermore, there is a risk that you will miss out on some useful research papers that have not been covered by a review article.

Select only more recent publications

Here you can choose what years' publications you want to consider. This allows you, for example, to narrow down to a single year if you wish. A rule of thumb for your assignment might be to narrow your search down to the last 5 years as a starting point. If we select only the last 5 years, then the number of articles from Web of Science drops from more than 8000 to 2300. That's still too many but it is a considerable improvement. Again there is a risk that you may exclude some older, but very useful papers.

Refine your search terms

'Muscle fatigue' is a very broad term to use for searching, so you could consider refining your strategy. If you enter the term 'muscle fatigue' into the search box, then the database will search for all articles that include the terms 'muscle' and 'fatigue' but this does not mean that the words will necessarily appear together as a specific phrase. If you want to search specifically for 'muscle fatigue' then, in Web of Science, you enclose the term in quotation marks:

"muscle fatigue"

This will now only select those papers where the two words appear together. Doing that reduces the number of articles selected from 8000 to 2500.

The title of the assignment actually referred to the 'mechanisms of muscle fatigue'. So a further refinement would be to include the word 'mechanisms', so your search term becomes:

"mechanisms of muscle fatigue"

This now reduces the number of articles to 13, all of which are directly relevant and will make a good starting point for the research.

This strategy worked fairly easily but you may find that not all searches are as simple and that you need to be still more sophisticated. A further refinement is to use what are called 'Boolean Operators'. The most common of these are AND, NOT, and OR.

These allow you to specify specific combinations of search terms. For example, if you specifically want to find out about muscle fatigue research that has been done on humans you could enter the following string:

> "muscle fatigue" AND human

This will now return a list of articles where the words muscle fatigue occur together and the word human also occurs. However, if you wanted to find out about any work that had been carried out on animals other than humans, you would enter the search string:

> "muscle fatigue" NOT human

Through these types of search operators, you can create very specific criteria for what articles you want.

Each time you undertake a search, you should look at the number of articles that the search has returned and scan the first few titles to check that they appear relevant. With a broad search you are likely to find a large number of articles, many of which are not directly relevant. As you refine your search, for example by selecting specific years or by using more specific search terms, then the number of articles should progressively reduce and should become more relevant, but there is a compromise as you may also start to exclude some articles that would be useful.

3.2.2 Web search engines

The major web search engines, such as Google, Google Scholar, Yahoo!, AltaVista, etc., are familiar to most people and they can be used to scan literally billions of pages of web-based information. They are very effective at producing long lists of websites related to your inquiry, but this means you have to be very selective in terms both of finding material that is directly relevant and then evaluating the reliability of that material: don't forget that anyone can host a site and put information onto the web and there is no guarantee that the information has been checked for accuracy.

Researching for our essay on the mechanisms of muscle fatigue using search engines, we can adopt the same procedures as when using the academic journal databases. If we start simply by entering the key words:

> mechanisms muscle fatigue

Google returns 1 700 000 results, Yahoo returns 615 000, and AltaVista returns 608 000, and all three engines ranked Wikipedia (see below) at or near the top of the list. The ranking of the list, however, does not necessarily indicate the direct relevance of the information to be found on the site (see Box 3.1).

BOX 3.1 Web search engines: a very brief introduction

When we are looking for information about muscle fatigue and enter those words into the search box of a search engine, the search engine will return sites that contain the words 'muscle fatigue', but the words will not necessarily be together, and in that order, unless you have used an advanced search strategy.

The order of the listing of the sites is also determined by a ranking process. This process ranks pages on a scale of 1–10 on the basis of the number of links to that specific site. The importance of each of those links is also scaled on the basis of the ranking of their sites. So the more links there are to a given web page and the higher the ranking of each of those sites, then the higher that web page will appear on the search engine's listing.

Although the high number of sites means that there's only time to check the first few, a scientific topic such as the mechanisms of muscle fatigue tends to come up with a significant number of research papers from academic journals and links to other academic sites such as universities in the first listings, so these are likely to be reliable sources for our essay.

If the title is on a more topical subject, the results are very different. For example, consider the 'effects of smoking' as a title. Using the keywords 'effects' and 'smoking', Google returned 15 200 000 results.

This means that, although the first pages in the listing will be popular, they are not necessarily the most relevant pages and you may need to scan quite a way through the listing to pick up the most relevant sites.

These links are to a whole variety of sites such as commercial sites offering products to help you stop smoking, government advisory sites, pressure groups (from various perspectives), media sites such as the BBC, charities and self-help organizations, medical advisory groups such as 'NHS Direct', and many more. You therefore have to exercise judgement in terms of the types of site you look at and you need to determine their reliability (see below).

It may be that, for your particular task, you do need a range of sources of information, not just research articles. You can therefore use the broad search engines and perhaps refine your search using the advanced search techniques available. As with the academic databases referred to earlier, these advanced search techniques allow you to specify search criteria, such as an exact phrase or without specific words, etc.

If you want to search for academic articles only, then the search engines now include sub-sections that will do that. An example is Google Scholar—a search engine that searches peer-reviewed papers, books, and conference proceedings. Although the range of journals that can be accessed by these types of searches is not as extensive as for the specifically academic engines, such as Web of Science, there has been a rapid expansion, and search engines such as Google Scholar do offer a good alternative.

How reliable is a website?

As we have seen, there are many sites providing information on specific topics and some are more reliable than others. So how do you know what to use? From a purely academic standpoint, the scientific journals, with their peer-reviewed papers, offer the most reliable sources of information, but there are many other useful sites as well. The simplest questions to ask yourself about each site are: who, or what organization wrote the material and how old is it?

You can quickly answer these questions by looking at the site address in the listing. Each site is identifiable by its URL (Universal Resource Locator). Examples of useful sites would include:

- *University sites.* These can include course materials from the university's under-graduate programmes, which can be very useful as an additional resource, as well as links to the publications and research outputs of their academic departments. Sites for higher education institutes are identifiable from the URL, which will contain the code ac (academic) or edu (educational). Thus universities in the UK all have an address that takes the form: www.[universityname].ac.uk. Those in the USA have the address www.[universityname].edu, and for Australia they are www.[universityname].edu.au.

- *Government sites.* These can provide information at a range of levels, particularly in terms of policy or public advice. Examples here would be the Food Standards Agency, the Environment Agency, or the National Health Service (NHS). Most government sites in the UK have the address form: www.[name].gov.uk, NHS sites have the address form: www.[name].nhs.uk.

- *Learned societies.* There is a learned society associated with almost every specific area within the biosciences, for example: Society for General Microbiology, Physiological Society, British Neuroscience Association, British Ecological Society, and many more. Most of these sites have addresses in the form: www.[name].org. Addresses ending in .org are used for organizations that are non-governmental and non-profit-making.

- *Professional bodies.* These are organizations that represent and maintain the pro-fessional standing of specific professions, such as the General Medical Council, the General Dental Council, and the Institute of Biology. These sites too usually have addresses as: www.[name].org.

- *Media organizations.* Many of these organizations, such as the BBC, *The Times*, *The Guardian*, and *The Daily Telegraph*, are very useful for providing information about topical science stories. Remember, however, that the story is usually written up by a journalist who may have a limited knowledge of the topic and is aimed at the general public rather than someone looking for detailed, scientific information. Media organ-izations are also in the business of selling news stories and, therefore, they may focus on the more sensational aspects of a topic.

- *Commercial organizations.* Many commercial organizations have very useful websites, but it is important to remember that they also want to sell a product. Going back to our search on the effects of smoking, there were several sites listed that were hosted by commercial organizations selling products to assist in stopping smoking. These carried quite a lot of useful educational information—for example, descriptions of the dangerous effects of smoking on the lung and cardiovascular system. However, they also carried a lot of commercial information setting out the benefits of their specific product. Although the information published by such sites has to comply with the requirements of the Advertising Standards Agency, they do not have to be objective in their approach: for example, while they are extolling the benefits of their own product, they are under no obligation to inform you that another company's product may actually be more effective.

- *Specific societies and organizations.* There is a huge variety of organizations and groups with websites, and the quality of these varies greatly. In the area of health, there are numerous self-help organizations that have sites that carry useful information, such as the Parkinsons' Disease Society, the Alzheimer's Society, or the Motor Neuron Disease Association. In environmental biology, there are organizations such as the World Wildlife Fund and Greenpeace. These sites are typically aimed at a non-academic readership but often carry a significant amount of background information. However, these organizations frequently have very focused agendas and therefore the information provided may be selective or interpreted in very specific ways—so when you are reading, you need to be aware of that focus and to maintain your critical approach.

Mention must also be made of one of the most used sources of information on the web, Wikipedia. Wikipedia is ostensibly an online encyclopaedia but it differs from all other encyclopaedias in that anyone can add or delete information from specific entries. This means that, whereas much of the information present may have been written by experts and should be accurate and current, an entry may also have been written by someone with very limited knowledge and may contain major inaccuracies. Unless you already know about the topic, you will not know whether the content is accurate or not. Wikipedia may have uses, for example for reminding yourself quickly about a topic, but it is certainly not advisable to use it if you are learning about a topic from scratch or researching for an essay. Your lecturers are unlikely to be impressed if the reading list for your assignment contains articles from Wikipedia.

3.3 Note-making strategies

It's past midnight and you are finally in the process of writing your essay. It's due in for tomorrow morning but that's okay because you have done all the research, you have your plan in front of you and the writing is going well. Then you hit a problem: you just need to describe the key experiment that supported the argument you are putting forward. You know you read up on it but you can't quite remember the details. It's somewhere in that pile of review articles but you can't remember which one and you don't have any notes to jog your memory. . . .

Situations like this are very frustrating but they can easily be avoided if you are systematic in making notes when you read up on a topic. Many students only think of note-making in the context of lectures but it is just as important in terms of your reading around the subject. We can divide note-making into two processes: the first is generic and comprises the things you should be doing irrespective of why you are making the notes. This includes making sure that the information you are using is properly referenced so that you avoid plagiarism (Chapter 8). The second process varies depending on what you want to use the notes for, e.g. developing understanding of a topic, revision purposes, or information for a coursework assignment.

In the following sections, we consider each of these in turn.

3.3.1 Generic note-making

To ensure that your notes are as useful as possible and that your note-making is done effectively, you must always record the source and make sure that your notes are notes.

Record the source

Always make sure that you record the source of the information. You never know when you may need to find it again: particularly when writing your essay in the early hours of the morning . . . the more details you give, the easier life will be later on (Figure 3.2). For example, just jotting down the title of a book is of limited help, particularly if it is several hundred pages long! Don't forget also that if you are note-making for an assignment, you will need the full reference for your reference list.

FIGURE 3.2 Reference details for notes from a book and a research paper. It's okay to use abbreviations in the details since you can go back to the full reference if you need it for your assignment.

Dimmock, N et al (2001) Intro to Modern Virology.
  Chap 16, p244
      Requirements of vaccines are ...
Tisdale, M (2000) Monitoring of viral susceptibility.
Reviews in Med Virol 10, 45–55
        p50 devel of 'flu inhibitors ...

Notes are notes!

You should always aim to note the information in your own words: just copying a paragraph out, or photocopying it, is not note-making. The process of reading and then summarizing the information in your own words is a very useful exercise because:

- as an active process, it will help you retain the information in your memory;
- it also tests your understanding of the information;
- it makes it faster for you to retrieve the key information at a later date, e.g. when revising.

Have a look at the extract from the sample text in Figure 3.3 and then the notes in Figure 3.4 taken from the sample text. Identify some of the techniques used to set out the notes. These include:

- use of highlighting to emphasize the topic;
- leaving space to keep the notes clear;
- leaving a wide margin, so you can add additional notes or explanatory comments later on;
- the source of the material (including page number) is given at the start of the notes;
- lines are used to indicate linkages of ideas and the division of the types of disorder into two groupings;
- the writing is in a very abbreviated style: there is no need to write in full sentences or include words like 'the', etc.

FIGURE 3.3 From Levy, A. and Lightman, S. 1997. *Endocrinology*, p. 91. Oxford: Oxford University Press.

Overactivity

The causes of overactivity of the adrenal cortex can be divided into intrinsic adrenal cortical problems, such as the formation of hormone-secreting tumours, and extrinsic problems such as excessive production of ACTH in Cushing's disease or the formation of renin-secreting tumours of the kidneys (Table 4.7).

Although almost all causes of adrenal cortical overactivity are rare or very rare, Conn's syndrome and congenital adrenal hyperplasia are probably both more common than is generally recognized. An overnight dexamethasone suppression test (see p. 372) is almost always sufficient to exclude Cushing's syndrome in patients with hirsutism or obesity.

FIGURE 3.4 Example of notes taken from a piece of text.

3.3.2 **Focused note-making**

As with your reading goals, the way in which you make notes will depend to some extent on the reason why you are making the notes. Some possible examples are:

- to aid with revision;
- as a tool to help with learning about a new topic;
- to aid understanding of a topic;
- as part of the research for a coursework assignment.

Assuming that you will be adopting the generic approaches described above, how do these goals differ in terms of note-making? You could look at them in terms of your requirements. For example, the first two goals are likely to be fairly broad in coverage of a topic, whereas the last two are likely to be more focused on a specific aspect of a topic.

Note-making to aid with revision

You are preparing for an exam and have a large volume of notes from lectures, some background notes from reading textbooks, and perhaps some notes from research papers. It is a very useful exercise to bring these notes together so that they form a coherent theme that you can read through, rather than chunks of information dotted around (Figure 3.5). As well as making the notes easier to revise from, the process of actually undertaking such an exercise also helps you to learn. This topic is addressed in more detail in Chapter 13.

Note-making as a tool to help with learning about a new topic

It may be that you have to research a topic from scratch rather than using note-making as a supplement for your lecture notes or handouts. In this case, the best approach is to build up your notes in stages, rather like the revision notes. To start, you need to get an overview of the topic, for example by reading and making notes from a general textbook (see goal 1, p. 24). Then go into the topic more deeply using a review article or research papers to focus on specific aspects. Finally, produce a set of composite notes by combining the overview notes with the more detailed ones. By the time you have done that, you should have a useful set of notes but should also have gained a good appreciation of the topic. Furthermore, when you come to produce the composite notes, it will also become clear to you if there are still areas that you do not understand.

Note-making to aid understanding of a topic

This task is a more focused note-making activity since you are likely to be concerned with a small aspect of a topic that you do not understand. In this case, your reading will be very goal-directed (goal 2, p. 24) and you only need to make short notes to supplement your current notes from your lectures. In this context, it is very helpful if you have large margins or space in your lecture notes (Chapter 2, *Making the most of lectures*) because then you may be able to insert the additional information within the original notes rather than having to write it separately. Amalgamating notes from different

FIGURE 3.5 Construction of revision notes from lecture notes and notes from a text.

| Muscle Organization | Dr Bond | Lecture 7 | 23/10/07 |

Motor units – functional grouping of m fibres
MU = 1 α motor neuron + group of mfs.
Size of MUs and contract. properties varies – 3 main groups – S, FR + FF
Postural muscles, e.g. soleus – mainly S
                             Muscles for jumping etc. –
gastroc. – mainly FF

Motor units
Human Physiology, Davies et al, p205
S Units – slow, weak contractions, oxidative metabolism, don't fatigue
FR Units – fast, intermed. contractions, ox/glycolytic metab, fatigue resistant
FF Units – fast, strong contractions, glycolytic metab, fatigue quickly

---

| Muscle Organization | Revision notes |

Motor unit – functional grouping of muscle fibres
Motor unit = α motor neuron + the group of muscle fibres it innervates
MU size + contractile properties vary
3 main groups of MU: S – slow contracting, weak, oxidative metab.,
       v. fatigue resistant
                    FR – fast contracting, stronger,
                        oxidative/glycolytic
                        Fatigue resistant
                  FF – fast contracting, strong, glycolytic,
                      fatigue quickly
Postural muscles, e.g. soleus – mainly S
Muscles for jumping etc. – gastrocnemius – mainly FF

sources is a very useful exercise, as discussed above, but it is not necessarily such a good use of your time if there are only very small sections of additional information that need to be included.

Note-making as part of the research for a coursework assignment

As part of the preparation for your assignment, you should have drawn up a list of specific topics that you need to address (see, for example, the section on *Make a plan* in

Chapter 5, *Writing essays*). Under each of those headings, you can now set out the key pieces of information from your reading, which will supplement other sources such as your lectures. As for understanding the topic (above), each point is likely to be quite brief and you will probably only need to record three pieces of information:

- the factual information itself;
- a summary of the experimental approach used to find the information;
- the full citation for the reference (see section 3.4 below).

3.4 Citations and references

When you are writing or presenting any piece of work, it is very important that you acknowledge the sources of the information and ideas that you are presenting. If you present an idea or a piece of information without acknowledging a specific source, then you are effectively claiming ownership of that idea or information. Provided it is your idea then that's okay: for example, it could be your interpretation of the results obtained in a practical class or from your research project. The failure to acknowledge sources, however, is one of the most common reasons for being accused of plagiarism (see Chapter 8, *Avoiding plagiarism*).

In this section we will look at a standard way of referencing and discuss when and how you should do it.

3.4.1 Defining the terms

The process of acknowledging sources sometimes seems more complicated than it need be because of the different terms used to describe what you should do. We will work through some examples but it is useful to explain some of the terms first.

Referencing is the name given to the process of acknowledging any material in the form of ideas, specific information, drawings or illustrations, experimental methods, computer programs, etc., that has come from someone else's work. Normally this is work that has been published in some form, for example as a research paper, book, or film, but it may also be simply a theory that someone has told you.

Citations are the specific acknowledgements to other people made, for example, in the text of your writing. Normally, these will be the name of the person and the date when their material was published: see some of the examples below in the short section 3.4.4 on *How to draft citations and reference lists*.

References are the list of sources that you actually referred to in your work. Every citation in the text should appear in the reference list and every source listed in the references should refer to a citation in the text.

Bibliographies are lists of all the sources that you made use of during your research. A bibliography may, therefore, include sources such as general textbooks that you used for background reading but from which you did not actually take specific material for your work. Some of the material listed in the bibliography may not, therefore, appear as a citation in the text. For most work at university level in the biosciences, particularly in the later years of your course, you will be expected to produce a reference list rather than a bibliography.

3.4.2 Why should I cite other people's work?

The main reason, as indicated in the introduction to this section, is so that someone reading your work will know where the ideas you are presenting came from and will be able to identify which ideas or information that you are presenting are specifically your own. The references are also very helpful to the reader who needs to know more about the background to the topic: he or she can go to the articles you have cited to find out more detailed information. When you are preparing coursework, the list of references will show your tutor what materials you have used and he or she may be able to give you feedback about the suitability of the materials and your search strategies.

3.4.3 Do I have to cite everything?

The general principle is that anything that you include in your work that is taken from someone else must be acknowledged. This is true just as much if you are quoting someone word for word, or if you are taking their ideas and putting them in your own words. Having said that, there is a level of information that is now accepted knowledge and for which it is not necessary to provide a citation. For example, if you write the statements:

> Chlorophyll is the main pigment involved in photosynthesis in green plants.
>
> Bats are mammals that can fly.
>
> Haemoglobin is the respiratory pigment involved in oxygen transport in humans.

These are accepted facts and there is no need to cite the source of the information. Indeed, the derivation of such information is often unclear anyway! To get a feel for how much you should cite, one of the best things you can do is think about it when you are doing your background reading: when you read a review article or a research paper, notice the level of information that the authors are taking as accepted and the levels at which they make specific citations. Have a look at the guidelines below as a framework for knowing when to cite your sources.

You should always cite the source when:

- you are quoting word-for-word from a source, whether it is a textbook, research paper, lecture, or film. Any such quotation must also be written in quotation marks (see below);

- you are copying, or taking elements from, a figure from a piece of published work;

- you are presenting information or ideas that are critical to the arguments you are making;

- you are presenting any ideas that may be controversial, for example someone's theory about a specific process;

- you are presenting information based on specific experimental or observational evidence.

If in doubt, then acknowledge the source.

3.4.4 **How to draft citations and reference lists**

If you look through different academic journals and, particularly if you look at publications in different subject areas, you will see that the citations and reference lists are formulated in a variety of ways. There are several standard styles for referencing such as Harvard, Vancouver, Chicago, and so on. Although the styles are different for these, the basic information is always the same. In this context, Harvard is one of the most common formats for citations and reference lists in the biosciences, so this is the format we will employ in this text.

In the Harvard system, the citation in the text simply comprises the name(s) of the author(s) and the date. In the reference list you must also include the author's (or authors') initial(s), the title of the article and the journal or book it comes from, the volume number, and the page number. In the examples below we will run through the most common variations for citations and references. Note that, even within the Harvard system, there are differences in presentation. The formats given below are based on the British Standard but you should check with your lecturers whether your department recommends any specific styles.

Referencing journal articles

For much of your study, articles published in scientific journals will represent the major source of information. Note that the formats for referencing given below apply to all journal articles, even if you downloaded the article from the journal website.

Citation:

If there are up to three authors for a given publication, then all the authors' names are given:

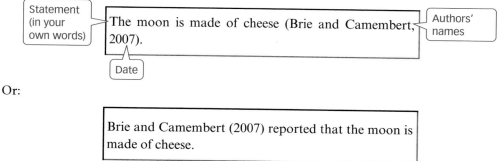

Or:

Brie and Camembert (2007) reported that the moon is made of cheese.

Reference:

If there are four or more authors, only the first name is given in the citation followed by *et al.* (this comes from the Latin expression *et alii*, meaning *and others*) but all the authors are listed in the reference.

Citation:

> The moon is made of cheese (Brie *et al.*, 2007).

Or:

> Brie *et al.* (2007) reported that the moon is made of cheese.

When giving a direct statement, as above, some authors prefer to use a phrase such as:

> Brie and co-workers (2007) reported that . . .

To cite two different papers by different authors in relation to the same statement:

> The moon is made of cheese (Brie and Camembert, 2007; Cheddar, 2008).

Or:

> Brie and Camembert (2007) and Cheddar (2008) reported that . . .

If you have more than two papers by different authors, listing the authors and dates in brackets is usually preferable to a long list of names in the body of the text. Note that the publications are normally listed in chronological order.

To cite two papers by the same author(s) in different years, list the papers in chronological order:

> The moon is made of cheese (Brie and Camembert, 2005, 2007).

To cite two papers by the same author(s) in the same year, the papers should be identified as 'a' or 'b' and this distinction added to the reference list (see below).

Citation:

> The moon is made of cheese (Brie and Camembert, 2007a,b).

References:

> Brie, A.B. and Camembert, C. 2007a. Structure of the moon: reflections on edible compositions. *Journal of Lunar Sciences*, **13**(3), 29–35.
>
> Brie, A.B. and Camembert, C. 2007b. Variations in the density of cheese on the lunar surface. *Journal of Lunar Sciences*, **14**(1), 54–63.

The assumption so far is that these journal articles are articles that you have read, i.e. that these are **primary** sources. It may be that you wish to cite the work of a researcher but have not read the material in the original article—for example, if you have not been able to obtain a copy of the article. In this case, you may have found that the information you need is quoted in a more recent paper by another author, and so you can refer to this more recent paper. This is clearly more indirect than taking your information directly from the original research paper, and it is referred to as a **secondary** source. The use of such secondary sources is not ideal, but sometimes it is unavoidable, particularly when you cannot access the original article. The format for such referencing is as follows (note that the reference is for the source from which you extracted the information, not the original source).

Citation:

> Cheddar had originally proposed that the composition of the moon was of blue stilton (Cheddar, 1964 as cited in Brie and Camembert, 2007).

Reference:

> Brie, A.B. and Camembert, C. 2007. Structure of the moon: reflections on edible compositions. *Journal of Lunar Sciences*, **13**(3), 29–35.

Referencing books

If you wish to cite material from a book, the citation in the text takes the same formats as for the journal articles: the name(s) of the author(s) and the date of publication.

The reference list format varies according to the type of book. The basic format is as follows:

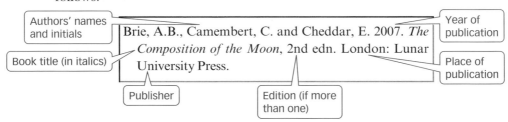

Authors' names and initials | Book title (in italics) | Year of publication | Place of publication | Publisher | Edition (if more than one)

Brie, A.B., Camembert, C. and Cheddar, E. 2007. *The Composition of the Moon*, 2nd edn. London: Lunar University Press.

Many books are compositions of articles that have been edited by the main editors. In this case, the reference would be:

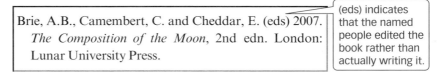

Brie, A.B., Camembert, C. and Cheddar, E. (eds) 2007. *The Composition of the Moon*, 2nd edn. London: Lunar University Press.

(eds) indicates that the named people edited the book rather than actually writing it.

In the case of such edited books, you may have only cited a specific chapter in the book that was written by another author. The format for such a reference is as follows:

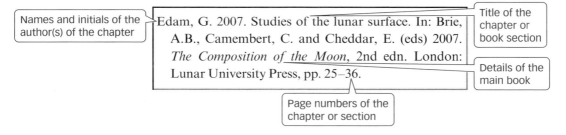

Names and initials of the author(s) of the chapter | Title of the chapter or book section | Details of the main book | Page numbers of the chapter or section

Edam, G. 2007. Studies of the lunar surface. In: Brie, A.B., Camembert, C. and Cheddar, E. (eds) 2007. *The Composition of the Moon*, 2nd edn. London: Lunar University Press, pp. 25–36.

Referencing websites

An increasing amount of information is available from websites and many such sites can be very valuable sources (though see section 3.2). As with any other source of information, that source has to be acknowledged through a citation in the text and a reference in the reference list. The citation in the text is, as for the other types of source, author(s) name(s) and date of publication. The basic format for the reference list is as follows:

Names and initials of the author(s) or organization name. | Type of medium [in square brackets] | Publication date | Article title (in italics) | Web address | Date the site was accessed [in square brackets]

Lunar Development Agency. 2008. *Mining the Lunar Surface* [online] Available from: http://www.lunardevelopment.co.uk/htm [Accessed 01 April 2008].

In this section, we have looked at the styles of citation and referencing for the most common sources of information that you might be using for your work. As stated before, there are different formats for presenting such information and here we have concentrated on the Harvard format, which is one of the standard forms used in the biosciences. You should, however, ensure that you check with your university or college to see if they have a preferred style of referencing.

✳ Chapter summary

In this chapter we have looked at the range of ways of working with sources to find information. This started with looking at the different types of sources available to you, ranging from textbooks through to research papers, and considering the type of information each sources provides and the situations in which you might make use of them. We then reviewed the processes of searching for information, using a variety of online approaches with a discussion of how to refine your searches and also how to assess their reliability. In the final sections of the chapter we considered the processes of making notes from these different sources and how to reference the material correctly in your written work.

Chapter 4

Choosing the right writing style

⊃ Introduction

Writing for scientific purposes is often very different from the styles of writing we may use for other forms of communication, as it normally requires use of a formal, impersonal style. It is important to use words accurately and avoid slang, informal phrases, and generalities. When we are talking with each other, we can use gestures to add to the meaning of what we say and we can repeat things or state them differently if we realize the listener has not understood us. It is also very easy to be rather lazy in our expression because we can see whether our listener has understood, '. . . know what I mean?' When you are writing, you have only the written words to convey your meaning and you cannot ask the reader whether they have understood or not, so precision is very important.

4.1 **Voice**

The style for scientific writing is normally impersonal and so avoids the use of personal pronouns such as 'I' and 'we' and the expression of subjective viewpoints. It also commonly involves using the *passive* voice rather than the *active* voice. For example, an introduction to your essay could be written as:

For a formal, scientific style it would be better to write:

Passive voice

> In this essay, comparisons will be drawn between the peripheral nervous system and the endocrine system . . .

You should also avoid using contractions of words and expressing a subjective view-point. For example, the conclusion to an essay might be written:

> The evidence I've given supports my feeling that . . .

This phrase contains the personal pronoun 'I' in the contraction 'I've'. It may be that, for particular reasons of emphasis, you wish to keep the word 'I' in this phrase, but you should avoid using any of the many contractions that appear in conversation such as 'I've', 'don't', 'couldn't', etc. Instead you should write them out in full: 'I have', 'do not', 'could not' (in this book we have chosen to use contractions because we feel that a more informal style is appropriate here, but it's (there we go again!) not appropriate for an academic essay).

The phrase also expresses a very personal, subjective viewpoint: 'my feeling'. In the formal voice, this could be written as:

> The evidence presented supports the argument that . . .

If it is important to express a personal summing up of the arguments presented, then you could write:

> On the basis of the evidence presented, I would conclude that . . .

The use of generalizations often clouds the meaning of what is written or indicates a lack of confidence on the part of the writer. For example, what does the following phrase actually mean?

> The peripheral nervous system quite often responds to stimuli . . .

The phrase 'quite often' has no real meaning as its interpretation depends entirely on the reader. Relative terms such as 'quite' or 'fairly', or phrases such as 'In general terms . . .' lack precision and should, wherever possible, be replaced with more precise phrases. Indeed, if you can, you should provide quantification of the statement, so rather than stating 'quite often' you can write:

> In 75% of cases the response was . . .

As with the choice of words, precision in grammar is very important to ensure that our writing is clear and unambiguous. For example, the incorrect position of a comma in a sentence can change the whole meaning of that sentence as in the old joke about the panda that walks into a bar, eats something, pulls out a gun and shoots the bar tender before leaving the bar. When questioned, the panda pulls out a book and points to the description of the panda which states it '... eats, shoots and leaves'. Had the description been phrased correctly as '... eats shoots and leaves', the bar tender would not have been shot! There are numerous, entire books covering the subject of grammar (for example, *Eats, Shoots and Leaves*, by Lynne Truss) so in this brief section we will look at just a few of the key rules.

Having read this section you should now be well aware that we have not written this book in a scientific style!

4.2 Sentences and phrases

The fundamental unit of writing is the sentence. Sentences may be short. If too many short sentences follow on from each other, however, the style of writing can seem very abrupt and disjointed, but overly long sentences, particularly if they contain several, inter-linked phrases, that might be associated with different points, can also be problematic because they can cause the reader to lose the thread of the argument being presented and give rise to the impression that the writer does not have a clear perspective of what he or she is trying to say.

The section you have just waded through is clearly badly written, and you may well not be sure what we were trying to say (as a short exercise, you could try rephrasing the paragraph so that the meaning is clear). So what is a sentence? At its most fundamental, a sentence comprises the following features:

> A noun is a word that identifies a person, a place or a thing

- a noun, that is the subject of the sentence;
- a verb;
- the beginning of the sentence is marked by a capital letter;

> A verb is a word that identifies an action

- the end of the sentence is marked by a full stop, a question mark, or an exclamation mark.

For example:

In fact, this sentence contains two nouns: 'Neurons' and 'cells'. The noun 'Neurons' is the **subject** of the sentence, and the noun 'cells' is used to give the reader more information about what neurons are. Although it is very brief, it is a complete sentence, whereas:

> Neurons are cells that are located

is a phrase not a complete sentence. Although it contains a noun and a verb, the sentence is incomplete because the verb 'located' needs to be qualified by a description of where that location is:

> Neurons are cells that are located in the nervous system.

is now a complete sentence. We may also include **adjectives** and **adverbs** in order to give greater detail to our descriptions to the reader. Adjectives are words used to describe a noun and adverbs are used to describe a verb. For example, the following sentence:

> The hormone, testosterone, is active in the body.

is a complete sentence but it does not give the reader very much information. By using adjectives and adverbs we can amplify the statement as follows:

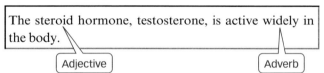

In the sentence above, the adjective 'steroid' tells the reader what type of hormone testosterone is, and the adverb 'widely' indicates to the reader the range of activity within the body (as a rough guide, adverbs usually end in 'ly'), though you should note the comments above regarding the use of relative terms.

BOX 4.1 Adjectives and adverbs

Adjectives are descriptive words used to describe a noun: 'a *brown* coat'. Adverbs are also descriptive and used to describe a verb: 'he ran *quickly*'.

If we wish to add still more information about testosterone, we could add a separate sentence:

> The steroid hormone, testosterone, is active widely in the body. It promotes the development of lean muscle mass.

In the second sentence, 'It' is the subject of the sentence but refers back to the previous sentence describing testosterone. As we mentioned earlier, too many short sentences make the writing disjointed, so we can link these two sentences together:

> The steroid hormone, testosterone, is active widely in the body and promotes the development of lean muscle mass.

Here, the phrase 'promotes the development of lean muscle mass' is giving a more detailed description about the role of testosterone. Note that this is now a **phrase**, not a sentence, because there is no subject within the phrase itself. The phrase is joined to the rest of the sentence by the **conjunction** 'and'. Conjunctions are words used to link phrases together, the most common being 'and' and 'but'.

With the aid of conjunctions, we can link many phrases together and end up with very long complex sentences which become difficult for the reader to follow. On the other hand, as we have already mentioned, a series of very short sentences is also hard to read. So, how long should a sentence be? There is no specific ruling on this but there are some rough guidelines you can follow. Most of the sentences in this section of the book are between 15 and 25 words in length. If your sentences are approaching 35 or 40 words in length (or more!) and comprise several phrases, you should think about dividing them up. Very brief sentences may be useful occasionally, for example to emphasize a point. If, however, you find you have written a series of sentences, each of which is less than 10 words in length, then see if the text will flow better if you combine some of them. The best test of your writing is to read it out loud. If you find it flows well and you can breathe in logical places, then you are probably about right.

4.3 Punctuation

Punctuation often causes students significant problems and written work may end up with commas, semi-colons and other symbols scattered around seemingly at random. As we saw with the case of the unfortunate bar tender being shot by the panda, however, the presence or absence of a well-placed comma can completely change the sense of a sentence.

4.3.1 The full stop

The full stop is probably the easiest form of punctuation to deal with. Full stops are used to indicate the end of a sentence and are followed by a capital letter which marks the start of the first word of the next sentence. Full stops are also traditionally used to indicate the use of a common abbreviation such as 'etc.' for *etcetera*, 'i.e.' for *id est* (meaning, that is), and 'e.g.' for *exempli gratia* (literally meaning, free example). In these cases, the full stop is not automatically followed by a capital letter. This usage of full stops for common abbreviations is decreasing and many writers simply put 'eg'.

4.3.2 **The colon**

The colon is most commonly used in the following ways:

- as in the previous phrase, to introduce a list or an example: 'A typical use of the colon would be as follows: . . .';

- to introduce a quotation: 'In his lecture, the professor stated that: "the nervous system . . ." ';

- to link two contrasting statements that might otherwise be written as separate sentences: 'The peripheral nervous system conveys information rapidly: this contrasts with the endocrine system . . .'. Here these two statements could have been written as separate sentences, but the use of the colon emphasizes the contrast between them.

- to provide justification for a statement: 'The peripheral nervous system conveys information rapidly: this is especially important when the body is responding to damaging stimuli'.

4.3.3 **The comma**

The comma is the most widespread form of punctuation and is frequently misused. Commas are used primarily as a means of providing a pause in the flow of the text; the pause can be in a variety of contexts, which are probably best illustrated by examples:

Separating items in a list

Commas can be used to separate items in a list: 'Oestrogen, progesterone, follicule-stimulating hormone, and oxytocin are all hormones that . . .'.

Separating out distinct phrases

Commas are used to separate out distinct phrases or clauses within a sentence. For example, we might write the sentence: 'Testosterone, which is one of the sex hormones, plays a role in the development of lean muscle mass.' In this context, the sentence would still be correct and have meaning if the clause 'which is one of the sex hormones' was not there. However, its inclusion adds further qualification to the statement.

The presence or absence of commas in the sentence can also change the meaning of that sentence. For example, compare the following two sentences:

> The adrenal hormones which are associated with the control of blood pressure, act on both the heart and blood vessels.
>
> The adrenal hormones, which are associated with the control of blood pressure, act on both the heart and blood vessels.

These two sentences have subtle but important differences in meaning. In the first sentence, the implication is that, of the hormones released by the adrenal gland, only those that are associated with the control of blood pressure act on both the heart and blood vessels. By contrast, the second sentence implies that all the adrenal hormones are associated with the control of blood pressure and act on both the heart and blood vessels.

Creating pauses

Commas are used to create a pause after the opening phrase of a sentence: 'In this essay, comparisons will be drawn between . . .' or 'As was stated earlier, the endocrine system is involved . . .'.

This is by no means an exhaustive list of the usage of the comma and we would encourage you to follow this up by reference to the more detailed expositions found in texts on English grammar, some examples of which are given in the further reading. As we have seen, the use of commas is important in imposing the correct structure and meaning to the sentences we write. To check whether you have placed your commas correctly, try reading aloud what you have written and ask yourself if the pauses indicated by the commas feel as though they are in the right place. This is by no means infallible, but it will give you a useful guide.

4.3.4 **The semi-colon**

The final, common form of punctuation is the semi-colon. Semi-colons are employed much less frequently than commas. The semi-colon can be thought of as being half-way between a comma and a full stop. It is typically used in the following ways.

- To separate items in a list where the sentence structure is complicated and commas are already in use to separate qualifying phrases, for example:

> The hormones that regulate blood pressure include: adrenalin, which is released by the adrenal glands and increases heart rate and peripheral resistance; angiotensin, which is released by the kidneys and . . .

- To link two related sentences together rather than separate them with a full stop, for example:

> Adrenalin is released by the adrenal glands; it has to be transported in the bloodstream to the heart.

4.3.5 **The apostrophe**

The apostrophe needs special mention in any consideration of punctuation because it is probably the most abused of any piece of punctuation, although it is actually fairly straightforward to use. Increasingly, the apostrophe can be seen randomly used before or after the letter 's' when the 's' is simply indicating a plural. This is incorrect. Apostrophes have two main functions: to indicate possession and to indicate contraction.

Possession

To indicate possession the apostrophe is added to the word followed by the letter 's', thus: 'The heart's blood supply is . . .'. The test here is whether the sentence could be rephrased using 'of': 'The blood supply of the heart . . .'. If the word is a plural, already ending in 's', for example 'The kidneys', then the apostrophe is placed after the 's': 'The kidneys' blood supply is . . .' (do the test: 'The blood supply of the kidneys is . . .'). If the word is singular but already ends in an 's', as is seen with some names such as 'James' then you still obey the rule for the singular form and add both an apostrophe and the 's', thus: 'James's heart condition worsened . . .'.

Contraction

An apostrophe can be used to indicate a contraction where some letters have been omitted because two words have been joined together to make a shortened form, such as 'don't' for 'do not' (remembering, of course, the earlier point in this chapter that you should normally avoid such contractions in academic writing!).

There are, however, some traps for the unwary:

it's who's

These both indicate contractions: ' it is' and 'who is'.

Its whose

These are both possessive pronouns: 'The body regulates its blood pressure . . .'. Other possessive pronouns that don't take the apostrophe are: his, hers, ours, and yours.

4.4 **Paragraphs**

Solid pages of unbroken text are very off-putting and are difficult to read, so the text is usually broken up by separating it into paragraphs. A paragraph is usually made up of several sentences that are linked together by a common theme. The break between paragraphs can then be used to separate out different sets of arguments to help structure the logical flow of the essay as a whole. For example, if you want to make four main points in your essay, then give each point its own paragraph. A possible paragraph plan might look like this:

- introduce the main idea (topic sentence);
- explain the idea (amplify the topic sentence);
- present supporting evidence or examples (quotation, study, expert opinion, or report);
- comment on the evidence (show how it relates to the main idea);
- conclude the main idea (link to the title or link to the next point).

A brief sample paragraph from our essay on the peripheral nervous and endocrine systems could be structured as follows.

Introductory sentence, setting out the theme of the paragraph	The speeds of communication of information are very different for the peripheral nervous system (PNS) and the endocrine system. In the case of the PNS, the conduction velocity of the action potential along the individual nerve axons can range from speeds of $0.5\ ms^{-1}$ up to $120\ ms^{-1}$.
Evidence to support the statement	The differences in speed depend on the axonal diameter and presence or absence of myelin, with the highest speeds being generated by the mechanism of saltatory conduction of the impulses along large-diameter, myelinated axons. The transport of hormones takes place through the vascular circulation. Here the speed of flow also varies, but in the arteries it ranges from 0.02 to $0.2\ ms^{-1}$. The differences in
Final sentence linking back to the introductory sentence	communication speed between the PNS and the endocrine system are important considerations when comparing the functional roles of the two systems.

There is no rule that prescribes how many sentences make up a paragraph, but the point where you are moving from one topic to another within the essay makes a logical place to insert a break and start a new paragraph.

4.5 Abbreviations

Abbreviations are often used in scientific writing, usually to remove the effort from writing out lengthy terms, in full, each time they are used. The usual rule of thumb is that you should write out the term in full, followed by the abbreviation in brackets, the first time you use it; on subsequent occasions you can just use the abbreviation:

The electrocardiogram (ECG) was recorded while the patient. . . . The ECG recordings showed that . . .

In scientific convention, some abbreviations such as 'ATP' or 'DNA' are so commonplace that they are often used without previously being written out in full. However, the safest practice is always to write the term out in full first.

Writing style exercise

Now that you have read through these brief sections on punctuation, try using the rules to punctuate the following section of text. (Before you criticize us for not including any references in this piece: the references have been deliberately left out to simplify the exercise).

> The peripheral nervous system comprises neurons whose cell bodies are located in the ventral horn of the spinal cord for the motor neurons and the dorsal root ganglia for the sensory neurons the axons of the motor neurons leave the cord via the ventral root and join the axons of the sensory neurons to form a mixed nerve these nerves branch along their route from the spinal cord to the tissues and end by innervating the peripheral structures such as the muscles and skin some sensory axons terminate as free endings within the muscles and skin whereas others innervate specialized sensory end organs in the skin these organs include the Merkel disks Pacinian corpuscles and Ruffini endings in the skeletal muscles the large diameter sensory axons innervate specialized proprioceptors called muscle spindles and tendon organs the axons of the motor neurons innervate the muscle fibres of the skeletal muscles with each motor axon innervating several muscle fibres but each muscle fibre only being innervated by a single motor axon the axons terminate in a specialized structure termed the neuromuscular junction action potentials are conducted along the axon until they reach the pre-synaptic end of the axon here the action potential causes the release of the neurotransmitter acetylcholine (ACh) that diffuses across the synaptic cleft and binds with receptors on the post-synaptic membrane the binding of the neurotransmitter causes the muscle fibre to depolarize and fire its own action potential which results in muscle contraction.

A suggested model for drafting this piece can be found at the end of the chapter, but try your own version first.

4.6 **Illustrations**

You should always think about using illustrations in your essays and other pieces of coursework since they can often be used to help you explain a process or describe a structure much more effectively than by using words only. Use of illustrations might include:

* flow charts to indicate the operation of a control system as shown in Figure 4.1;
* diagrams to help explain a life cycle or a biochemical process;
* drawings to help describe anatomical structures or the features of an organism;
* drawings of apparatus described in practical reports.

There are some key points to remember when including illustrations.

- Produce your own illustrations, don't copy and paste a diagram from a website: don't forget that the rules of plagiarism apply as much to drawings as to the written word (see Chapter 8, *Avoiding plagiarism*). Also, just copying a diagram will not help you to understand it.

- Give each illustration a figure number and refer to it from the text, don't just leave it as an 'add on'.

- Give each illustration a title so that the reader can see what is being shown.

- Make sure that your drawings are labelled so that the reader can identify the different structures being shown.

- Feel free to use computer drawing packages or freehand for producing your drawings: don't forget that in examinations you will probably have to draw your diagrams by hand.

Remember also that illustrations can also save you words, so if you are struggling to write your essay within the word count, then a well-chosen illustration can save a lot of words in description or explanation.

Figure number

Explanatory figure title

FIGURE 4.1 Flow chart illustrating the feedback control of the release of growth hormone by the pituitary. GHRH, growth hormone releasing hormone. (Adapted from Berne and Levy (2000), Figure 44-8.)

Citation for the source of the figure

Abbreviations are spelled out in the legend

Explanation of symbols

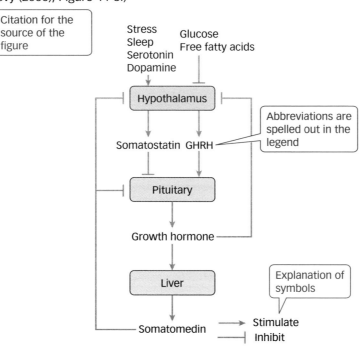

4.7 Referencing

Finally, in this chapter on choosing the right writing style, we need to flag up referencing. We dealt with this topic extensively in Chapter 3, *Working with different information sources*, and we will return to it again in Chapter 8, *Avoiding plagiarism*; however, it deserves a brief mention here because referencing is an important part of academic writing style. You must ensure that any information you get from another source is appropriately referenced.

✳ Chapter summary

Writing for a scientific audience requires a very formal style, with precise use of language. In this chapter we have looked at the conventions of writing style such as the use of the passive form and an impersonal, rather than personal, expression. We have also revised some of the principles of basic grammar and construction. Finally we have highlighted the importance of referencing, more of which can be found in Chapter 3.

📖 Further reading

Seely, J. 2007. *The Oxford A–Z of Grammar and Punctuation*. Oxford: Oxford University Press.

Truss, L. 2007. *Eats, Shoots and Leaves: The Zero Tolerance Approach to Punctuation*. London: Profile.

Writing style exercise: suggested structure

The peripheral nervous system comprises neurons whose cell bodies are located in the ventral horn of the spinal cord for the motor neurons, and the dorsal root ganglia for the sensory neurons. The axons of the motor neurons leave the cord via the ventral root and join the axons of the sensory neurons to form a mixed nerve. These nerves branch along their route from the spinal cord to the tissues and end by innervating the peripheral structures, such as the muscles and skin. Some sensory axons terminate as free endings within the muscles and skin, whereas others innervate specialized sensory end organs in the skin; these organs include the Merkel disks, Pacinian corpuscles, and Ruffini endings. In the skeletal muscles, the large diameter sensory axons innervate specialized proprioceptors, called muscle spindles and tendon organs.

The axons of the motor neurons innervate the muscle fibres of the skeletal muscles, with each motor axon innervating several muscle fibres but each muscle fibre only being innervated by a single motor axon. The axons terminate in a specialized structure, termed the neuromuscular junction. Action potentials are conducted along the axon until they reach the pre-synaptic end of the axon; here the action potential causes the release of the neurotransmitter acetylcholine (ACh), that diffuses across the synaptic cleft and binds with receptors on the post-synaptic membrane. The binding of the neurotransmitter causes the muscle fibre to depolarize and fire its own action potential, which results in muscle contraction.

Chapter 5

Writing essays

➜ Introduction

Essay-writing is a very common form of assessment in almost all degree courses in the biosciences, both as coursework and for examinations. For many students, however, writing essays at university represents a real challenge as it is a skill that is only developed to a limited extent in pre-university courses. It may also be a long time (perhaps years) since you last wrote an essay, and that might have been for an English course rather than a piece of scientific writing. This chapter is designed to help you appreciate what is expected of you, and guide you through the process of planning and writing essays. Before we begin though, consider this point: we often refer to the process of producing an essay as 'essay-writing', but how much of the time is actually spent writing?

Think back to the last coursework essays you produced: how much of the total production time was spent writing and what else was involved in the process? If you are like us, then your time allocation for essays is probably something like this:

- quite a few hours searching for information in textbooks and research papers and even more time reading through the material trying to understand it!

- following that, you have to decide and note down what is specifically relevant and should be included;

- then some sort of planning stage, where you decide how to order the material and how to begin and round-off the essay;

- finally, there is the writing itself, which probably only represents about 20% of the total time spent working on the essay.

5.1 **What is an essay?**

So exactly what is an essay? Dictionaries and web sites abound with definitions of essays, viewed from different perspectives, some being more helpful than others! In historical terms, the word essay means an attempt or trial. The *Oxford English Dictionary* includes the following definitions:

> The action or process of trying or testing.
>
> A composition of moderate length on any particular subject, or branch of a subject; originally implying want of finish, 'an irregular undigested piece' (J.), but now said of a composition more or less elaborate in style, though limited in range.

YourDictionary.com defines an essay as follows:

> A short literary composition on a single subject, usually presenting the personal view of the author.

One of the problems is that the definition depends on the subject area and the expectations of the readers. Perhaps the easiest approach is to consider what we might call the 'typical' essay in the biosciences. This would be a piece of structured writing, of 1000–3000 words in length, that is written as an answer to a specific question or in response to a specific title. In this context the student is expected to be able to display knowledge and understanding of the subject material, setting it out in a coherent manner and drawing logical conclusions from the material presented in order to answer the original question.

5.1.1 **What are essays supposed to achieve?**

This is probably a question many students have asked themselves, particularly during the early hours of the morning as a deadline is getting very close! It may also be a question that markers have asked themselves too, as they wade through a pile of exam scripts or coursework essays. Actually, a well-designed and well-answered essay question involves the development and testing of a significant number of skills, some of which are not explicitly expressed. It is all too easy to focus on the scientific content of the piece of writing and not to think much about the way that content is presented and structured, but the skill of writing is a critical element in the production of a good essay.

If we take the brief for a piece of coursework such as:

> In 1500 words, compare the endocrine and peripheral nervous systems as channels for control and communication within the body. The essay is to be submitted to Dr Harrison by 12 noon on Friday 1 March.

We can identify a list of the skills being exercised in order to complete the task as follows:

- using feedback from previously marked essays to produce a better piece of work;
- analysing the question;
- researching the subject matter;
- bringing together the information gathered from a range of sources, e.g. lecture notes, text books, research papers, web articles, etc.;
- planning the presentation of the information to address the question and make sure that all the key points are covered;
- ordering the material in a logical manner;
- framing the body of the text with an introduction, to set the scene, and a conclusion, to summarize the points made;
- writing succinctly and with clarity, paying attention to the grammar, syntax, and spelling;
- referencing the information correctly to acknowledge the original sources;
- proof-reading the essay to pick up on editorial mistakes and ensure clarity of writing;
- managing your time, so you are not scrabbling around trying to write the essay an hour before it is due in, even though you were given the title 4 weeks previously.

This is a surprisingly long list and it probably could be extended. The important point is that while some of these skills, such as researching the topic, are obvious and easily recognized as part of the process, others, such as using feedback and proof-reading, are more implicit but are still very important if you are to write a good essay. As we go through the whole process of preparing and writing the essay, we will consider these different skills. Other aspects, such as researching the material, are covered in Chapter 3, writing style is covered in Chapter 4, and using feedback is discussed in Chapter 12.

5.1.2 Principles of scientific writing

Before we can get to grips with answering the essay question comparing the endocrine and peripheral nervous systems, we need to consider some of the features of scientific writing. Most people can remember the very first essays they wrote at school, probably on the theme of 'What I did in my holidays' or something similar. These essays are usually written in an informal style and referenced very much to the individual (Figure 5.1).

FIGURE 5.1 Your first essay (probably!).

It was a lovely hot day and I went to the beach with my sister. We fought a lot and Mummy got cross.

Writing for a scientific audience, whether it is an essay, a report, or a research paper, is a very different exercise and requires a different, and more formal, style, as we explained in Chapter 4, *Choosing the right writing style*.

5.2 Approaches to essay-writing

Producing a coursework essay involves a number of key stages, including:

- analysing the question/title;
- researching the topic;
- making a plan;
- writing a first draft;
- reviewing and re-drafting;
- proof-reading and checking citations and references.

Ideally, these stages would be allocated appropriate portions of time to ensure that the production of the essay is as smooth and straightforward as possible, as represented by the timeline in Figure 5.2. The division of time won't be precise and doesn't need to be the same for each essay, but you do need to allocate time for all the important elements that make up the production of an essay, not just, for example, the writing. Importantly the timeline also includes a gap between writing a first draft and writing a second draft. This will ensure that you can be more objective in your review and make significant improvements to the work.

FIGURE 5.2 Ideal allocation of time for essay-writing tasks.

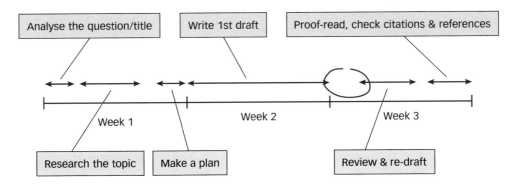

5.2.1 Common problems

The above represents an ideal approach and we would encourage you to aim to allocate time to your essay-writing tasks in a similar fashion. If you can allocate your time in this way it is a reliable method for increasing your chances of completing an essay on time and to an appropriate standard. However, in practice, the process of producing an essay is rarely as smooth and straightforward as this. Some common problems faced by students (and academics) when writing are as follows:

- leaving things to the last minute;
- doing too much research;
- doing little or no planning;
- taking too much time over the writing.

All of these are real problems, many of which you may have experienced for yourself. Each of the problems has a range of possible causes, and what causes the problem will determine what you need to do to address the problem. We will deal with each in turn.

At the last minute

You might find yourself trying to complete your essay at the last minute because:

- you have lots of other deadlines to meet;
- you feel anxious about writing essays and so avoid starting them;
- you just struggle to summon up the motivation to start earlier.

If you do write an essay like this it means that you can be very focused and the work is often completed in the minimum time (because it has to be!). However, it can be a very stressful way to write and will affect the quality of your work because you run out of time to review it and so end up handing in what is effectively only a first draft. If you find yourself in this situation your immediate priority is to organize your time more effectively (as shown Figure 5.2) so that you can allocate time to improving your essay by reviewing and re-drafting.

Too much research

You might find yourself undertaking too much research because:

- you get side-tracked by information that is interesting but not relevant;
- you do a lot of reading which generates a lot of notes;
- you make very detailed notes on what you read.

If you do write an essay like this it means that you have lots of material for your essay and you feel well informed; however, the volume of information can be overwhelming, not all of it is relevant and there is less time available for other stages of the essay-writing process. Researching the topic is a vital task—nevertheless you don't get judged on the information you have gathered but on what you have written down. If you find yourself in this situation your immediate priority is to develop a clear focus for your reading and note-making (see Chapter 3, *Working with different information sources*). This will make you more efficient in gathering information, so you can then allocate sufficient time to the other stages. You may also be spending too long on research because you are not very effective at finding the material you need. Again, have a look at Chapter 3 for some tips about search strategies.

Little or no planning

You might find yourself doing little or no planning because:

- you run out of time (see *At the last minute* above);
- you find planning a bit restrictive and you would rather just start writing because writing helps you to think.

If you write an essay like this because the writing process helps you to think, it will probably mean that your writing will flow well and so it will read quite fluently. However, it can be a confusing and inefficient way to write because it is difficult to get an overview of the material, and the logical sequence of the material can suffer as a result. If you find yourself in this situation your immediate priority is to set aside time for planning. The aim of planning is to provide you with an overview of the essay content so you can select and order your points in the most effective way. The essay plan can be in linear note form or it may be in the form of a visual plan (see section 5.3.4), but whatever the format, it will provide you with instructions for selecting and ordering the content of your essay.

Writing takes too long

You might find yourself taking too long over the writing because:

- you re-write individual paragraphs again and again;
- you write too much, then struggle to cut it down to the word count;
- you simply find getting the words down on paper very difficult.

If you do write an essay like this it means that the writing gets a lot of attention, which is a good thing. However, it can be an inefficient and frustrating way to write and won't leave you with enough time for other aspects of the process. If you find yourself in this situation your immediate priority is to speed up the production of the first draft in order to have more time for reviewing and re-drafting.

In this case, you need to have a plan of action for the writing process. Start with a set of bullet points that summarize what you want to say in the essay, and then just write a paragraph about each point, allowing yourself an approximate number of words for each paragraph as a proportion of the essay as a whole. Don't try re-writing any sections until all the paragraphs are written. When you have the body of the essay in place, you can think about linking the paragraphs together. At this stage, you will have a good appreciation of how long the essay is going to be and also of the overall structure of the essay. Now you can devote some time to editing, or re-structuring the essay.

5.3 Producing a coursework essay

We can now use the steps identified in section 5.2 to show you how to produce a course-work essay, based on the following stages of production:

- plan your use of time;
- analyse the question or title;
- research the topic;
- make a plan;
- write the first draft;
- review and re-draft;
- use feedback effectively.

5.3.1 Plan your use of time

You will have noticed from the common problems identified above that crucial to improving your writing is managing your time. Many of us (the authors included, as we are sure the Editor would confirm!) are not very good at time management. When we are set an assignment and told that the deadline for submission is not for 6 weeks, the natural reaction is often to put it in a drawer and forget about it until a few days before the submission date. That means that many of us err towards being 'at the last minute-ers'. We all know that this can lead to problems.

- 'I've only 2 days in which to write the essay and have just found out I can't get hold of a copy of the textbook because other students have already borrowed it.'
- 'The library doesn't have access to the journal article I really need and I have to find something else. Help!'
- 'The essay for this module is due in at the end of the week but I've also got two other deadlines coming up . . .'

So when should you begin working on an essay assignment? The obvious answer is 'as soon as you are given the title and your instructions'. It may not be practicable to race off and start writing immediately, but you do need to incorporate your assignments into your course planning. Clearly, the sooner you start the better, but you may need to consider some other factors as well. For example, although you have to submit the essay in week 8 of the term, you know that you won't finish covering the topic in lectures until the end of week 4, so you are probably better off not trying to do anything until you have the background from the lectures. An example is shown in Figure 5.3.

One useful approach to your planning is to put all your assignments onto a single chart or diary. For example, you could use a calendar. When planning, it is best to work

FIGURE 5.3 Planning your time for an essay.

1	2	3	4	5	6	7
			Essay set	Analyse question		
8 Key lecture	9 Research topic	10	11	12	13	14
15 Planning	16	17 Writing	18	19	20	21
22	23 Reviewing & checking	24	25 Essay due!	26	27	28
29	30	31				

backwards from the submission date and break the assignment down into its separate elements, as described above. This also has the advantage of giving you positive feedback because you can reward yourself as you cross each section off on the calendar. If you make sure that you incorporate the submission dates for other pieces of work as well, then you can plan your time effectively and also avoid hitting an unexpected log-jam of work that needs doing and handing in all at the same time.

When planning your timeline, you need to be realistic about allowing yourself enough time for the different components of the work; for example, carrying out research can be very time-consuming, particularly if you need to search for and read through (and understand!) a number of research papers. By the same token, you also need to be strict with yourself when undertaking the research and not get carried away looking for more and more papers.

Obviously, you won't be spending all your waking hours during those 2 weeks working on the essay: you will have other commitments such as lectures and practical classes, as well as keeping time for socializing. Therefore, within your timeline, you will need to decide on what days you are going to focus on the essay and for how long. Again, the key features are that you should try to be realistic, and then do your best to stick to your schedule.

What happens if things go wrong? Advance planning is again the key: whenever possible, leave a gap between your planned completion of the work and the submission date, then you have some leeway if something unexpected throws your plans out of line.

5.3.2 **Analyse the question or title**

It is very important that you write the essay that is asked of you: there is no point in writing an essay that doesn't address the question asked, however brilliant that essay may be. So, you have your time plan in front of you, and are ready to go. STOP. Before you do anything else, read the question carefully and then read it again, highlighting the following features:

- the subject of the question or title;
- the instruction;
- the key aspect;
- other significant words.

Figures 5.4 and 5.5 show a couple of worked examples for you.

FIGURE 5.4 Analysing the question or title: example 1.

FIGURE 5.5 Analysing the question or title: example 2.

Sometimes it might seem that you are having to highlight almost every word, but it is still worth doing if it helps you register all the important pointers.

There are several standard instructions used for essay questions, some of which appear to be very similar but with some differences in emphasis. The most common instructions are:

- *Compare*—is asking you to identify and comment particularly on features in common but also aspects of difference between two or more identified elements;
- *Contrast*—this is also a form of comparison but focusing predominantly on the areas of difference. This may be included with compare, as in: 'Compare and contrast . . .' thereby emphasizing that both the similarities and the differences must be brought out with equal weighting;
- *Describe*—is asking you to write a detailed description of the subject;
- *Discuss*—is often used to mean the same as describe, but it can also be used in its more literal sense of discussing the merits or shortcomings of a particular statement or idea, as in the example given above (Figure 5.5);
- *Explain*—is often used in the context of making clear how or why something happens, for example: 'explain the process whereby the blink reflex habituates during repeated stimulation';
- *Evaluate*—the strict interpretation implies some form of quantitative assessment of information but this may commonly be used in the more qualitative context of evaluating an idea;
- *Give an account of*—this usually has the same meaning as describe;
- *Write an essay about*—again this typically means describe.

Note that the example in Figure 5.5 includes the word 'and'. This was to ensure that both sections of the title were addressed: 'Discuss . . . and describe . . .'. Although the first section may be the more important of the two, you need to give sufficient time and effort to the second aspect, even if it does not necessarily amount to being half of the essay. The pitfall to avoid is writing a very long first section and then suddenly realizing that you still have the second aspect to deal with, so that it is tacked on, very clearly as an afterthought.

Once you have identified what basic type of essay you need to write, check for some other key pointers that give you some specific details about the approach to be adopted. These may include the following:

- *Write an illustrated account* . . . this is emphasizing that you need to include diagrams or other forms of illustration;
- *Using named examples* . . . make sure that you use specific examples that illustrate your arguments;
- *With reference to the recent experimental evidence, discuss the process* . . . ;
- *With reference to the research literature, compare the model for . . . with. . . .*

These groups of titles clearly show the emphasis that the question setter wants you to take in writing your essay. For example, it is not just a factual description that is required for the last two examples: the essay should be heavily weighted towards using the experimental evidence or recent research papers to support your arguments.

Finally, make sure that you are very clear about the breadth of the subject material required in your essay: look for the key aspect that specifies a group or range. For example, in an essay on the endocrine system, the title might refer specifically to steroid hormones; a title about ecology might refer to a specific type of habitat; or a title about animal behaviour might refer to a specific group or groups of animals.

5.3.3 Research the topic

Researching the topic is probably the most time-consuming part of the whole exercise. Your lecturers will no doubt hope that you are really enthused by the research you undertake and that you explore the subject in depth. That's great, but you need a good dose of pragmatism in your approach and to be aware that undertaking the research could be literally an endless task. Therefore, you need to keep your timeline in mind and, while working as efficiently as possible, be very clear that you will need to set specific limits to the amount of research you undertake. You can also save yourself a lot of time by some initial investment: sit down with a blank sheet of paper and brainstorm some ideas—simply write down, in any order, all the ideas you have—about what types of information you need for the essay.

You will also need a strategy for your research: particularly when using online journals and other web-resource search engines, it is very easy to end up with very large numbers of hits, many of which are not directly relevant to your needs. See Chapter 3, *Working with different information sources*, for a discussion of approaches to searching.

Use appropriate sources

There is an increasing amount of material available for research purposes and you should be strategic and critical in the way in which you use these different resources. These resources are dealt with in detail in Chapter 3, but a summary is also given here. The main sources of information are listed below and it is probably a good strategy to work through them in this order so as to keep a clear focus. This approach also means you are starting with relatively simple, well-defined information and progressively moving to the more complex, wider sources.

- Lecture notes and handouts—these will often provide the starting point for your investigations, by reading through what your lecturers have already told you about the topic.

- Textbooks—you can use textbooks to expand your knowledge and understanding of the principles of the topic. If you are in the very early stages of your university career, textbooks may provide the bulk of the information you need for the essay.

- Reading lists—for many modules you will be given lists of suggested reading, that provide additional subject depth and breadth, to enable you to develop your understanding of topics covered in lecture courses. These lists will normally comprise journal articles and similar materials that have been selected carefully by your lecturers as being appropriate for the module you are studying. You may even be given some suggested titles as a starting point for your essay assignments.

- Research journals—as you progress through your course, there will normally be an increasing expectation that you will be reading and using research papers that you have found in research journals in order to develop the subject content of your essays. Through your library or computer service you will probably have access to useful search engines such as Ovid, Web of Science, and Medline, that you can use specifically for searching through the vast number of research journals that are available in electronic format. As we mentioned earlier, you will need to be careful in developing your search pattern (section 3.2.1). You can do this by selecting the appropriate key words and seeing how many hits you obtain. Gradually refine your keyword selection until you have a limited number of hits that should be closely relevant to the subject matter of your essay. You will probably still have several tens of articles in your list and reading all those in full would clearly be very time-consuming. Reading the titles of the papers your search has come up with is therefore the next stage of selection. When you have drawn up a manageable shortlist of the titles that appear most relevant, then you can go to the next stage of reading the abstracts and, from your final list, select those papers which you need to read in full (having first checked that you can get access to the full articles!).

- Web resources—there is a vast and ever-increasing volume of material available on the web and it varies in quality from the excellent to the downright wrong (section 3.2.2). When you are looking through the journals of professional scientific bodies, you have the confidence of knowing that the research papers have been reviewed by scientists who know the subject area very well. Therefore, you can be confident about the validity of the material being presented. By contrast, the material that is posted on websites may have been posted by experts in the field but it may also have been put there by people who have little or no real knowledge but who have views about a subject that they want to broadcast to the world. So, you need to be cautious and critical and look for specific signposts that can indicate the reliability of the information presented: for example, if the website is published by a university or a professional body, then the information is likely to be accurate and reliable. If it is published by a commercial organization, for example a drug company, it too is likely to be accurate but may also be subject to a certain amount of marketing 'spin', for example to encourage take-up of a specific drug. At the other end of the spectrum, if the material is published by an individual or organization with no obvious professional qualifications or academic affiliations, then you should treat the site with considerable caution. Mention here must also be made of 'Wikipedia'. Wikipedia is an online encyclopaedia to which anyone can contribute. It is continually increasing in volume but, because of its nature, the quality of the material is variable: some is of high quality and has been written by experts in their respective fields. Some is much less accurate and is not necessarily reliable. Some is wrong. Unless you are an expert who already knows the field, the problem you will have is that of picking out what is good from what is not.

Make relevant notes

You now have in front of you a large amount of information that you have gathered from your research. Most of it will be in the form of articles extracted from different sources. Some of these articles may be quite lengthy and all of them will contain more material than is needed for your essay. So the next stage is to extract the relevant sections from these articles and put them in note form in your own words. The section on *Notes are notes!* in Chapter 3 (p. 37) should guide you here.

5.3.4 **Make a plan**

There is often a great temptation, having done your background reading, simply to sit down and start writing. This is, generally speaking, not a good idea: it is very easy to drift off the topic of the essay, to leave important points out, or to produce a very disjointed essay. Planning is a very important part of essay-writing and will help you gain the best possible marks by allowing you to produce a coherent structure to your essay.

The exact timing of when you do the planning, however, will depend on your approach; there is more than one way of using plans in essay-writing. An essay plan can be used before or after the writing of the first draft, and your preference will depend on how you approach the task of writing essays.

As discussed earlier in this chapter, conventionally, the order would be:

- analyse the question or title;
- research the topic;
- make a plan;
- write a first draft;
- review and re-draft;
- proof-reading and checking citations and references.

But if you are one of those people who struggle to write an essay plan, what should you do? The answer to this question depends on why you struggle to plan. Consider the descriptions below and think which approach best describes your way of working.

Writer 1

I like to make plenty of notes before starting. I like to have a good idea of what I am going to write before I begin the first draft. I find it difficult to begin a draft if I am not clear on the direction my writing is going to take.

If this is you then you need to use essay plans before you begin the draft, as per the conventional approach.

Writer 2

I tend to make fewer notes than most people and tend to want to start writing as soon as possible. Often I don't know what I want to say before I start to write, but I find that ideas occur once I have started. For me, most of my thinking is done through the process of writing.

If this is you then you need to set an early deadline for the completion of the first draft so you have plenty of time to review and reorganize the content. This is important because the content of your first draft will only be organized in the order in which the ideas occurred during writing. You will therefore need to spend more time on reviewing the draft than a person who has planned the structure beforehand. Your approach to writing can be very effective, but you must allow lots of time to review and improve the structure of the first draft.

Writer 3

I like to make plenty of notes beforehand and probably have a general of idea of the essay content before I begin to write. Once I have started to write I find that new ideas occur to me so the content of my draft is often different from my original intentions.

If this is you then you need to use essay plans before you begin the draft, as per the conventional approach.

Let's go back to one of the essay titles we discussed earlier:

> Compare the endocrine and peripheral nervous systems as channels for control and communication within the body.

There are many different ways of planning your essays, so we will look at just two different formats: the linear plan and the visual plan.

Linear plan

An example of a linear plan is shown in Figure 5.6. This is very useful if you like producing lists of key headings, to which you can then add sub-headings and specific details. Having drawn up the list of key points, you can easily re-order them to create a logical structure for the essay. You can also see very quickly, by referring to your notes, if you have any gaps in your research that need some additional detail before you start writing. Furthermore, once you know how many key points you have, you can divide up your word allocation between the different points, with the target of giving a paragraph over to each point. When you do start writing, you can tick the points off as you go through them, and so make sure that you don't leave anything out. This approach also means you are not likely to drift away from the topic, which can happen if you just start writing without any planned sequence to the essay.

FIGURE 5.6 An example of a linear map for writing a course essay.

Introduction (approx 200 words)

Overviews of the endocrine and peripheral nervous systems
Descriptions of control – making things happen – and communication – sending information

Body of the essay (approx 1600 words)

- Mechanism of communication (200 words)
  - Endocrine – hormones transported via blood stream, slow, widespread action – e.g. thyroid hormone
  - PNS – action potentials conducted via axons, fast, specific to a locality
- Action at cellular level (300 words)
  - Endocrine – can be via specific receptors on cell surface triggering second messengers – e.g. adrenalin
  Or crosses the cell membrane to act directly on the nucleus, e.g. testosterone
  - PNS – acts via specific receptors on the cell membrane – causes a change in the electrical potential of the membrane – e.g. acetylcholine
    - Point 3 (300 words)
      ○ . . .
    - Point 4
      ○ . . .

Visual plan

An example of a visual plan for the same essay is illustrated in Figure 5.7. Here the plan takes the form of linked ideas that are gradually developed into more specific details. This is often a very useful approach if the essay is conceptually difficult to put together and you are not sure how you can structure it. Once you have your basic plan drawn, you can allocate the numbers of words to the different levels so that each major branch is given appropriate weight. Again, as you write, you can tick off the branches that you travel along. For example, starting from the centre, you can tackle communication as your first topic. So you can write a paragraph comparing the routes by which the communication occurs, namely the bloodstream and the nerve axons; then moving to the outer branches, you can compare the characteristics of each means of communication. When you have completed this topic, you return to the centre point and start along the next main branch.

FIGURE 5.7 Developing a visual plan for an essay.

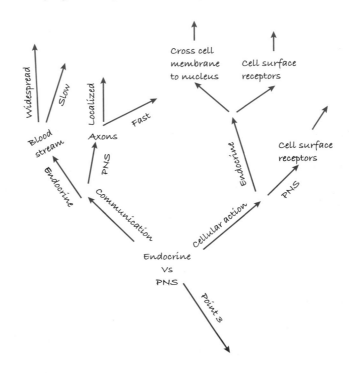

5.3.5 **Write a first draft**

There are three main sections to an essay that you will need to write:

- the introduction;
- the body of the essay;
- the conclusion.

We will look at writing each of these sections later in the chapter but it is useful to have in mind a structure for the essay before we begin trying to write. One way of summarizing what these sections comprise is: 'tell your readers what you're going to tell them, tell them (in detail), tell them what you have told them'.

The introduction to the essay is used to frame the question and to set out what aspects are going to be addressed. An introduction is typically one or two paragraphs in length and should certainly not comprise more than 10% of the essay word count. The body of the essay is the bulk of the essay. In this section you set out the arguments, in detail and in a logical order. Finally the conclusion is used to draw the reader back to the main focus of the question and to summarize the key points that have been made in the body of the text in a way that clearly addresses the question. As with the introduction, this should not normally exceed 10% of the essay.

Body of the essay

Regardless of when and how you do your essay plan, you should have gathered together from your research the evidence to support your points. The inclusion of this evidence is a very important part of scientific writing. It is not sufficient to state:

> The endocrine system has widespread actions on the cells and tissues of the body.

Each statement should be supported by a specific example, so:

> The endocrine system has widespread actions on the cells and tissues of the body. For example, the thyroid hormones act to stimulate the metabolic rate of all the body's organ systems (Berne and Levy, 2000).

Note that the statement of the example is supported by a citation, indicating where the idea you have described came from (see section 3.4 *Citations and references* for more information on referencing). As you write each section of the body of the essay, make sure that you mark off each point on the plan and then make sure that the paragraphs link together, rather than appearing as completely independent sets of statements.

If, for example, you want to make five main points in your essay, then give each point its own paragraph. A paragraph is usually made up of several sentences that are linked together by a common theme. The break between paragraphs can then be used to separate out different sets of arguments to help structure the logical flow of the essay as a whole. A possible plan for each paragraph might look like this:

- introduce the main idea (topic sentence);
- explain the idea (amplify the topic sentence);
- present supporting evidence or examples (quotation, study, expert opinion, or report);
- comment on the evidence (show how it relates to the main idea);
- conclude the main idea (link to the title or link to the next point).

Introduction

The introduction is a signpost for your reader, showing how you intend to answer the question. Remember what we said above about the structure of the essay: in the introduction you are going to tell the reader what the essay is going to cover. One possible structure for an introduction could be:

- begin with a general point about the central issue;
- use the words of the title to show how you will focus on the question;
- indicate what the structure of your writing will be;
- make a link to the first point.

Note that in our consideration of the writing process, we have placed the introduction after the main body of the text. There is no rule that states that you have to write the sections in the order in which they will appear in the final document. Some people find it is logical to write that way, but others prefer to write the introduction after the rest of the essay, because then they have a clear picture of what it is they are introducing.

Conclusion

The conclusion is where you summarize for the reader your main points, linking back to the question and highlighting the most important aspects of your material. It acts as another signpost to your reader, rounding off the essay. As such it gives you the opportunity to:

- refer back to the question to show that your essay directly addressed the set question;
- summarize briefly for the reader the main points your essay covered;
- show the overall significance of the material in relation to the question;
- provide an overall assessment of theories or arguments, summarizing your own viewpoint.

Again, you might want to write this in a different order from the final document: for example, you might like to have a conclusion written out first because you can use it as a guideline for reminding yourself where the essay is heading.

5.3.6 **Review and re-draft**

Reviewing and re-drafting are important stages of the whole production process, but they often get overlooked because not enough time has been left to do them effectively. Again, this is where planning your use of time is so important: you should aim to leave yourself time after writing the original draft to review what you have written rather than simply aiming to get the basic writing done by the deadline. Indeed, review is most effectively done when you have left a gap of a few days between writing the draft and undertaking the review. As identified in Figure 5.2, if you read through the essay immediately after you have written it, it is often difficult to look at it critically.

There are three key stages to the process of review:

1. Reading through the essay critically, ask yourself:
   - Does the essay answer the question?
   - Is there a logical flow from one paragraph to the next?
   - Does the introduction set the context of the essay and explain what is being written about?
   - Does the conclusion summarize the key points and set out the conclusions you have drawn?
   - Have I put in citations to support the key statements in the text?

- Are any diagrams or illustrations in the right place and have I included the source reference if necessary?
2. Re-drafting the essay to improve the structure following your review.
3. Proof-reading the essay to:
   - correct errors of spelling and grammar;
   - check the citations match the reference list.

Re-drafting

Assuming that you have written your first draft in good time, you may wish to re-draft the essay. The extent to which writers may wish to write more than one draft of an essay depends on how careful the initial writing stage was (as illustrated with the examples of writers 1, 2, and 3). Some writers prefer to get something down on paper (or computer) and then juggle around with it until they are happy with the structure; others work very closely to a structured plan, and therefore may only write a single version of the essay.

Re-drafting can take two stages: re-ordering of paragraphs to help improve the flow of the essay and then re-writing individual paragraphs to improve the way in which the points are conveyed to the reader. If you have written the essay using a word processor, then re-ordering is relatively straightforward by cutting and pasting the paragraphs into a different order. Likewise, with the individual paragraphs, it is easy to re-write a paragraph and insert it back into the text.

Don't forget to re-read the whole essay, to make sure that your re-ordering of the paragraphs hasn't broken the links between the different points.

Proof-reading

It is very tempting, when you have finished writing the last word of the essay, just to print it out and sit back feeling satisfied that you have finished and it is on time (maybe only just!). It is very important, however, that you proof-read the essay, checking for errors in your typing, spelling, and grammar. There is a danger that writers become too reliant on the spelling and grammar checkers in word-processing packages and assume that because the program has not highlighted a piece of text, then it is correct. These packages are not infallible. For example, many typographical errors will in fact result in a valid word so the spell-checker won't highlight it (for example 'to' and 'too' are both valid words but they clearly have different meanings). Conversely, many programs will flag up many of the technical terms you may use in the biosciences. For example, the word-processing program used for this book identified words such as 'angiotensin' and 'osteoblast' as being incorrectly spelt. As a consequence your essay on screen may be highlighted in lots of places and so there is a tendency to ignore all of these as 'false positives'.

How should you go about proof-reading? One of the hardest parts of proof-reading is that, when we read through something we have written, we read what we expect to see and can very easily overlook mistakes. This is especially true if you have just finished writing the essay and are still feeling totally immersed in the exercise. If you have managed your time carefully you will have at least a few days spare before you have to hand in the essay.

So, put the essay away for a few days and then sit down and read it carefully, as if it had been written by someone else and you were tasked with finding errors.

If you look to the punctuation section in Chapter 4, *Choosing the right writing style*, the best way of checking this is to read the essay out loud, as if to an audience. You might feel really stupid doing it, but you can always shut yourself away so no-one will know! As you read make sure that there are breaks in the text (and that you don't run out of breath!) and that they come in the right places for the sense of the text. Reading out loud is also a good test to make sure that the structure as a whole makes sense.

Another useful exercise is to put a guide (e.g. a ruler) underneath each line as you read through, so that you focus on the words you are reading. Doing that, you are much more likely to pick up on typographical errors that you would not spot otherwise (and which your spell-checker won't pick up either), such as 'to' instead of 'too', or 'lose' rather than 'loose'.

Finally check your use of references:

- make sure that your key statements are supported by citations in the text;
- check that each citation in the text is matched against a reference in the listing at the end of the essay;
- likewise, check that each reference at the end appears as a citation in the text.

Is proof-reading really that important? Yes, it is for a number of reasons:

- the person who is marking your essay will not be impressed by reading text that is littered with errors and, as a consequence, you will probably get lower marks than the scientific content alone might merit;
- some mark schemes include a specific allocation of marks for grammar and spelling;
- errors in punctuation and syntax can lead to ambiguities of meaning, so the sense of what you are writing is no longer clear;
- it is good to get into the habit of writing accurately: if you are asked by an employer to write a technical report, that employer will not be pleased if there are numerous mistakes;
- furthermore, when you are trying to sell yourself in job applications, these mistakes can be very costly: employers have been known to say that the first filter they use when processing applications is that any application with three or more spelling mistakes is automatically rejected!

So, get into good habits early and always check what you have written.

5.3.7 **Use feedback effectively**

It is very tempting, when receiving a piece of marked work, to look at the mark or marks obtained, feel moderately satisfied (or disappointed) and then put the essay away and not look at it again. It may well be that, since you submitted the essay, you have moved onto another module and don't feel that any comments would be relevant to what you are doing now. If you do this, you cannot hope to improve the way you write because you are not paying attention to the guidance given as to where you went wrong or how you could have done better. Likewise, it is also important to know what things you are doing particularly well, so you can continue doing them.

Using feedback is a very important part of the learning process and we go into more detail on this topic in Chapter 12. Feedback on essays should come in two forms: there should be subject-specific feedback, for example the identification of factual errors or omissions, and more generic feedback related to your writing style. The subject-specific aspects should help improve your understanding of the topic and may be particularly useful when you come to revise for examinations. The generic feedback should enable you to improve your essay-writing for the future, irrespective of the topic. This generic feedback might address aspects such as the use of references, the layout of the essay, or the use of illustrations.

So, before you consign your essay to the back of your cupboard, take time to read through it carefully: re-reading an essay after a few weeks is itself a good exercise because you will have a very different perspective compared with when you were immersed in writing it. Read through and think about the feedback you have been given. First of all make sure that you understand why the essay as a whole got such good or bad marks: was it focused on the question? Did you go into enough detail? Did the essay have a clear structure with each point being explained and then leading on to the next? If you don't understand where you went wrong, make an appointment to see your lecturer so you can talk through the aspects you don't understand.

A really good exercise is to keep a notebook in which you keep a record of the key points of feedback from your different pieces of work. In your book, note down the generic feedback from the current essay and then look back over the previous pieces of work to check whether this is an issue that has arisen before. If it is a problem that has been flagged up more than once before, you might well consider seeking specific advice about how to tackle it. Having noted down the key points, write a list of a few action points to bear in mind for the next essay you write, such as:

- write a more detailed plan;

- write shorter paragraphs;

- check the referencing style and that all references cited in the text are listed in the references.

Then, when you come to write your next essay, refer to your list of action points and make sure that you improve on those aspects. In this way, you should be able to maximize the benefit from the feedback given and progressively improve your writing. Don't forget, though, if you can't read the feedback comments or don't understand them, don't just give up: go and ask your lecturer for help.

5.4 **Writing essays in examinations**

Most bioscience courses at university will require you to write essays under examination conditions at some stage. In this section we are focusing on some tips in relation to essays; further guidance relating to examinations in general is given in Chapter 14, *Getting the most out of exams*.

So, you are sitting in the examination hall with the paper in front of you giving a list of titles. You don't have 6 weeks to think about, research, and write each essay: you now have only 3 hours and you must write three essays in that time. After the initial panic, try to be very methodical in your approach and follow these key points: 5 minutes of care at the beginning of the exam can save you a lot of heartache later.

Read the instructions on the paper through carefully and slowly, paying special heed to any that specify how many essays you should answer and whether you have to select titles from specific sections of the paper.

- Read the titles through carefully.

- Eliminate any titles about which you know nothing (hopefully not too many!).

- Select a set of titles on topics that you can answer, then narrow that down to the requisite number by checking the titles again, highlighting the key words, and then selecting those titles that you are confident you can address in full.

- Plan your first essay: again, time spent planning the essay and checking against the keywords in the title will help make sure that you answer the question.

- As you write the essay refer back to your plan and tick off the key points as you address them.

- Don't forget to include illustrations if they are relevant.

- Be strict with yourself about timing. It is very easy to select the first essay title as one you know most about and get carried away writing it, so that you don't leave yourself enough time to do justice to the other essays. So don't allow yourself to spend a long time on the first essay; divide your time up equally.

- Leave enough time at the end for reviewing.

If you only answer two essays instead of three, or select your titles from the wrong section of the paper, or don't address the question, you cannot hope to get good marks!

A final tip for this section: for your coursework essays you will probably have had a long time to write them, you will have written them using a computer, and probably did not write them at one sitting. Examination essays are very different: at the moment, most examination essays are written long-hand, using a pen. So, it's a very good idea to get some practice in writing under these conditions; this can form a valuable part of your revision. When you have finished revising a specific topic, look out some past examination papers and find a question on that topic. Then sit down and write an essay, under examination conditions. This will be a good test of how much you know and understand about the topic but it will also give you a feel for how much you can write in the time available. As a final guide, ask a lecturer to read through the essay and comment on it, so you can confirm that you are on the right track.

✳ Chapter summary

Essay-writing is an important skill, and, as with all skills, it develops with practice. In this chapter we have shown the importance of:

- planning your use of time as soon as you have the essay title so you don't end up rushing at the last minute;
- analysing the question or title carefully so you make sure you write the essay your tutors want;
- researching the topic carefully, using the different sources available to you;
- planning the structure and order of the essay;
- reviewing, re-drafting, and proof-reading to make sure you have answered the question, that the structure is logical, that the citations and references are correct, and that there are no spelling or grammatical mistakes.

Chapter 6

Writing practical reports

➔ Introduction

The knowledge base in all areas of the biosciences is underpinned by careful experimentation and observation. The abilities to design experiments, to undertake them methodically, to observe and record the outcomes accurately, and to analyse and interpret the findings are vital to the development of a scientist. These skills, however, are not sufficient on their own: you also need to be able to communicate your findings to other people. If someone asks you for directions to find somewhere, knowing where to go is only part of the issue; you also need to be able to communicate that knowledge. In the scientific community this is done in a number of ways, the most common being presentations at scientific meetings and published reports.

If you are planning to follow a career in the biosciences, you will need to develop all of these skills. However, such skills are not just of value in a scientific career: there are many careers where the skills of accurate recording, analysis, interpretation, and presentation of information are highly valued by employers.

During your studies you are likely to undertake a range of practical work, including practical laboratory work, fieldwork, and independent project work. The aims of these types of work are to develop your skills as a scientist and also to deepen your knowledge and understanding of the subject. The aims of this chapter are to help you get the most out of your practical work and to help you prepare high quality reports. Although we will discuss fundamental aspects of data presentation, this will be fairly limited since there are other guides that go into this in great detail. Likewise, we will not discuss the processes of experimental design or data analysis. Good guidance for these aspects can be found in one of the partner books in this series.*

*Holmes, D., Moody, P. and Dine, D. 2006. *Research Methods for the Biosciences*. Oxford: Oxford University Press.

This chapter is laid out so that you can read it straight through, or dip into sections for guidance on how to tackle specific aspects of your work. During your studies you will be expected to read scientific papers and it is always a good exercise not just to focus on the scientific content but also to evaluate the way in which the science is presented. As you read on, it will be helpful to have some research papers to hand so you can look at the different styles of presentation.

Although the focus of the chapter is on laboratory-based practical classes, much of the material is also directly relevant to fieldwork as well. To help create some context, the text uses as its examples an undergraduate investigation into the jumping ability of the locust. The chapter has been structured to follow the timeline of the stages of the practical class from initial preparation to completing the report, as shown in Figure 6.1.

FIGURE 6.1 Structure of the chapter.

6.1 **Before the practical class**

The writing up your practical work is the final stage of the process, and you can only hope to produce a good write-up if you have undertaken the work effectively in the first place. If you turn up to the practical class not knowing what you are supposed to be doing and not having checked that you have brought with you everything you need, then you are certainly in danger of working less efficiently in the class and are also likely to perform less well. Good preparation beforehand is therefore a key stage to success in all types of practical work.

Much of this section will seem to be simple common sense but if you want to get the most out of the practical work you are going to do, it is essential that you prepare properly. This is something that many of us are not very good at: it is very easy just to arrive at a class and hope everything will be made clear as you go along. However, not only will effective preparation help you learn more and get better marks, but it can often pay off in terms of saving you time and effort when you are actually doing the work. There are few things more frustrating than having spent a long time setting up an experiment only to discover that you have set it up incorrectly or not recorded all the data you need, so you have to start all over again.

Good preparation is particularly important if you are going to be working away from your normal site, for example in the field, when you cannot easily go back to fetch something you have forgotten. More specifically, good preparation means knowing:

- where you have to be and when;
- what items you need to bring with you;
- what you are going to be doing and why;
- some background to the investigation;
- how you are expected to present your findings.

For many practical exercises you will be given a schedule beforehand that provides most of this information; the following steps will also help.

6.1.1 Read the schedule

Reading the schedule carefully will enable you to check the time and location of the class and what items, such as a lab coat, calculator, etc., you need to take to the class. A lined notebook for recording your results and any notes about the experiment is always useful and much better than a borrowed sheet of file paper. It is good practice to get into the habit of writing your experimental notes in a notebook: this will be particularly important for a research project, and many scientific employers require you to keep detailed notes so that, if necessary, results can be verified at a later date.

6.1.2 Read the details of the experiment

Reading the details of the experiment will give you a good idea what you will be doing. It may well be that the techniques and equipment will be new to you, but familiarizing yourself with the description of the procedures will help when it comes to doing the experiment. Highlight any key sections of the methods, for example if specific measurements need to be made at specific times: Figure 6.2 shows an example of how you might do this for an experiment. If you are going to be working in a group, getting together to plan how you are going to organize yourselves and divide up the workload will make you more efficient and save time.

FIGURE 6.2 Using highlights to identify key parts of the experimental protocol.

Jumping Distance: experimental measurements
Measurements must be made in a warm environment (if possible 28–35°C).

You should try to obtain records from at least ten adult locusts. Each locust should be weighed and the length of the femur of the hindleg measured.

Take care not to mix up your locusts during the recordings: use ink-marks on the hindlegs for identification. The wings of the locusts should be fixed together with a small strip of adhesive tape to prevent them from flying.

6.1.3 Read up on the background

Reading up on the background to the work will help you understand the biology involved and can put what you are doing in context. It will help to read through your notes from related lectures. If there is a reading list, try to read through at least some of the material beforehand and make notes of key points (see *Note-making strategies*, section 3.3). There is also the bonus that, if the papers or texts are not available online and need to be borrowed from the library, you won't have to wait because other people have borrowed them first!

6.1.4 Check if there are instructions about presenting your results

Checking how you should present your results is important because it may vary from class to class. Sometimes you will simply be required to fill in tables or draw graphs on the schedule itself and answer some specific questions, other times you will need to write a full report similar to a scientific paper. Again, knowing in advance what is expected of you will help you to be more efficient in your work and save you time in the long run.

6.2 During the practical class

If you have done some preparation for the practical class it is much more likely that the class itself will go well. There are two important things you need to do during the class to ensure you can write your report afterwards: make notes and record your results.

6.2.1 **Make notes**

Try to get into the habit of making notes during your practical classes. Many practical classes will begin with a briefing session addressing a variety of aspects of the work, such as safety issues and instructions and demonstrations of the techniques to be used, as well as some background subject information. You may find you are having to stand up to see and hear what is going on, but, even so, do take notes because otherwise you will forget everything very quickly. There will be other times during the practical class when it will be useful to make notes as well—for example, if you need to make changes to the procedures during the course of the experiment.

Before beginning the experiment, check through the sequence of tasks that you have to complete and, especially for timed measurements, make sure you have a clear list of what needs doing and when.

At the end of the class, you may well have another briefing session. Make the most of the information you are given because this could be very important for when you are writing up: make good notes—don't just rely on your memory!

6.2.2 **Record your results**

Recording your results is clearly a critical part of undertaking any piece of practical work. Increasingly you will find that you are recording experimental data directly into a spreadsheet or other program. But whether you are storing the information in a spreadsheet or in your laboratory notebook, you still need to draw up tables for the results, making sure you identify what is being measured and what each column of the table represents.

During the experiment record your results carefully, clearly identifying numerical values and the units in which they were measured. Also make notes of any changes you made to the protocol, or to the way in which the results were measured. Again, care taken at this stage will save you a lot of time and frustration later on. You are also much less likely to make mistakes. Figure 6.3 illustrates two very different ways of recording the results from a simple practical class measuring the distance locusts can jump.

Compare the two sets of records in Figure 6.3 and list the aspects that you think are unclear in the data sets in sample B. If you were presented with these two data sets, would you be able to write about them in a meaningful way? Certainly one advantage of recording your results directly into a computer is that you won't have to try to decipher a colleague's handwriting (or even your own!).

6.2.3 **When things go wrong**

What if things go wrong? By their very nature, experiments may go wrong and you find you don't have any data or that they do not make sense. If this happens speak to your tutor as soon as possible—it is likely that you will be able to obtain another set of results from your tutor or from your colleagues in the class that you can then use for your report.

FIGURE 6.3 Two very different ways of recording your experimental data.

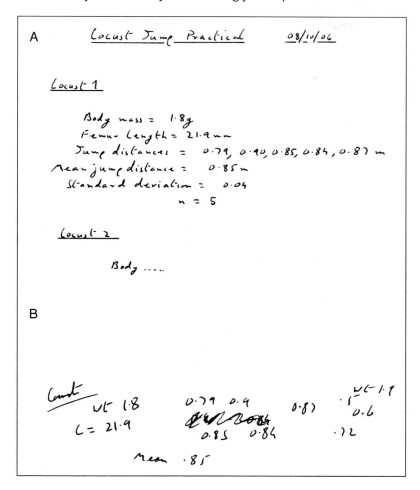

6.3 Writing your report

During your degree programme you will probably be asked to write up your practical work in a range of different formats. These may range from simply filling in tables and writing short answers to questions listed on the practical schedule you have been given, to preparing a full report in the form of a scientific paper. In this section we will take the scientific paper as the model, as it covers the most common aspects of report writing. Make sure you are clear what format of report is expected from you before you write!

6.3.1 **Use an appropriate writing style**

The principles of writing for a practical report are the same as those for other pieces of scientific writing and have been discussed in detail in Chapter 4, *Choosing the right writing style*. The key point to remember is that you are writing for a professional readership and therefore you need to write in a formal, scientific style.

Avoid the use of 'I' or 'We'

So the phrase: 'I measured the locust jumps at ten minute intervals . . .' becomes: 'The locust jumps were measured at 10 minute intervals . . .'. Likewise: 'On the basis of these findings, we concluded that . . .' becomes: 'On the basis of these findings, it was concluded that . . .'. This is known as writing in the **passive voice**.

Write in the past tense

You should describe what *was* done, or what *was* concluded. The only exceptions to this are when you are discussing illustrations or data presented in the report and the conclusions you can draw from them. For example: 'Figure 1 *shows* that jump distance *increased* with body mass . . .' or 'The key conclusion of this report *is* that jump distance *increases* with body mass . . .'.

Reference your sources

Wherever you take other people's ideas, you must reference them correctly in the text and list the sources in the reference list at the end of the report. Remember that not to do so is plagiarism (Chapter 8, *Avoiding plagiarism*). Referencing is dealt with in detail in Chapter 3, *Working with different information sources*.

Use SI units

Measurements should normally be presented using the SI (Système International d'Unités) conventional form. The three fundamental units are the kilogram (kg), metre (m), and second (s). The SI system of units is based on seven 'base units' which are the fundamental measurements from which the other units are derived (Table 6.1). From these base units

TABLE 6.1 The seven base units of the Système Internationale d'Unités (SI units)

Quantity	Unit	Symbol
Length	metre	m
Mass	kilogram	kg
Time	second	s
Temperature	Kelvin	K
Amount	mole	mol
Current	Ampere	A
Luminosity	candela	cd

TABLE 6.2 Common units and their definitions

Measure	Unit	Symbol	Derivation
Area	square metre	m^2	
Volume	cubic metre	m^3	
Volume	litre	l	$0.001\ m^3$ $1\ dm^3$ (1 cubic decimetre)
Velocity	metre per second	$m{\cdot}s^{-1}$	
Acceleration	metre per second per second	$m{\cdot}s^{-2}$	
Force	Newton	N	$kg{\cdot}m{\cdot}s^{-2}$
Energy, work	Joule	J	$N{\cdot}m$ $kg{\cdot}m^2{\cdot}s^{-2}$
Power	Watt	W	$J{\cdot}s^{-1}$ $kg{\cdot}m^2{\cdot}s^{-3}$
Concentration	mole per cubic metre	$mol{\cdot}m^{-3}$	
Temperature	degrees Celsius	°C	K
Electric potential difference	Volt	V	$W{\cdot}A^{-1}$
Electric resistance	Ohm	Ω	$V{\cdot}A^{-1}$

we derive a series of units for different measures. Table 6.2 lists some of the most common units that you may come across in practical work and their definitions.

Note that when writing compound expressions, e.g. $m{\cdot}s^{-1}$, you should use the middle dot · to separate the terms. This avoids confusion where terms could be misinterpreted; otherwise, for example, $N{\cdot}m$ could be mistaken for nm (nanometre) when hand-written.

6.3.2 **Get the structure right**

The standard format of a scientific paper is as follows:

- Title
- Abstract
- Introduction
- Methods
- Results
- Discussion
- References

You may also wish to include some additional sections such as Acknowledgements, Appendices (see *Results*), and Keywords or a list of abbreviations used.

We will look at each of these sections in turn and you may find it useful to refer to some of the papers from your reading lists to look at the way these sections are organized and presented.

Title

The title must be brief and informative, so that the reader can see at once what the paper is about. In the case of scientific papers for publication, the title is very important as it will be used as part of the information for electronic searching, so authors will often include keywords to increase the probability of the paper being picked up by an electronic search.

Often journals will limit the number of characters that can be used in the title. A typical limit would be 120 characters, including spaces, so it is good practice to write your titles within this limit.

Have a look at the four sample titles below, all of which describe the same investigation.

1. An investigation into the relationship between the body mass of an adult locust, of the species *Schistocerca gregaria*, and the maximum distance the animal can jump under non-restrained conditions.

2. The relationship between body mass and maximum jump distance for the adult locust, *Schistocerca gregaria*.

3. The locust jump.

4. Is bigger really better? Testing the jumping ability of different sizes of locust.

Which title provides the key information most efficiently?

The first title is unnecessarily long at 200 characters, though it does contain all the essential information. Here, the first section of the title 'An investigation into . . .' (or equally 'A study of . . .') is redundant and can be deleted without changing the sense of the title. The third title is too brief since it does not give any indication of the aspect of the jump being investigated. The second title is 108 characters long and contains all the essential detail for the reader to know whether to read further or not. Slightly humorous titles, such as the fourth example, may well be appealing for certain reports, and are often styles used by the popular press, but this approach should be treated with caution!

Abstract

There is a real art to writing a good abstract. The abstract must provide the reader with all the essential information from the paper within a very limited number of words, typical abstracts being around 150–200 words in length. In many cases, the abstract may be used as a free-standing source of information: for example, when you are researching for an essay, you may well only read the abstract of a paper rather than the whole paper.

The abstract will normally be the last part of the paper that you write and it is worth going through the following checklist to make sure you have covered everything.

- Background: very brief but enough to set the context for the . . .
- Aims: the key question you set out to answer using the . . .

- Methods: again these should be described very briefly, summarizing the technique(s) used to obtain the . . .

- Results: here you need to quote the key findings of the investigation, including the actual values obtained from which you drew the . . .

- Conclusions: the key points you drew from the results. The conclusions should be related back to the initial aims.

The results and conclusions are the most important part of the abstract because, in a scientific paper, this is where you are presenting new findings. These two aspects should therefore make up the bulk of the abstract, with the background, aims, and methods being written as succinctly as possible. When you read through the abstract of your report (or better still ask a friend to read it through), the two points that should be immediately obvious are: what was found and why it is significant.

Compare the information content of the two sample extracts from abstracts describing our locust jump experiment and identify the key pieces of information that are missing from the second extract.

Abstract 1:

The capacity of adult locusts to jump long distances to avoid predation is well known; however, it was unclear whether there is a relationship between the body mass of the adult locusts and their overall jumping performance. To investigate this, 20 male, adult *Schistocerca gregaria* (mean mass 1.6 ± 0.3 g) were stimulated to jump by means of a standard visual stimulus and the mean distance measured for ten jumps. A minimum of 15 minutes was allowed between each stimulus presentation to avoid habituation. The overall mean distance for the jumps was 0.95 ± 0.07 m and there was no correlation between body mass and mean jump distance ($r^2 = 0.15$). The findings indicate . . .

Abstract 2:

Adult locusts can jump long distances. In these experiments the jumping ability of adult locusts was measured. The adults all jumped in response to a stimulus and the jump distances were averaged. Sufficient time was left between the jumps so that the response did not habituate. The results did not show any relationship between the mass of the locusts and the distance they jumped. The findings indicate . . .

Although the second abstract does describe the conclusion of the experiment, it does not give any indication of the initial aims. Nor does it provide any of the key information and numerical data to allow the reader to appreciate what was done and what was found.

As with the title, the abstract of a paper is commonly used as a means of electronic searching and so should contain keywords that identify the nature of the study and its conclusions.

It is not normal practice to include references within the abstract.

Introduction

The introduction is the section of the report in which you describe the background and aims of the study. You might like to think of this section as addressing four questions.

- What has been done before?
- What still needs to be resolved?
- Why is it important to resolve this question?
- What do I aim to do?

A useful approach to drafting the Introduction is to start by setting out the aims of your study ('what do I hope to find out?') as a series of bullet points. Keep this list of aims in front of you while you are researching and writing the background to the study so that you keep clearly focused on the topic and don't start writing about sideline issues.

For a report in the form of a scientific paper you will need to read published papers on the topic of your investigation so that you can write a brief overview of the research that is underpinning your study. The skill in drafting this part of the introduction lies in writing a succinct summary of the findings of the previous authors and of the experimental evidence supporting those findings. As with all such writing, it is important that you present the material in your own words and fully reference the statements (see *Citations and references* section 3.4).

The background text is normally presented as a factual account without discussion of the ideas presented in those papers. In most areas of research, there are areas of knowledge that are accepted as agreed and which you can describe as the background to the study. Developing from these, there will be areas where there is still disagreement, or questions remaining to be answered. In the case of disagreements in the published research, you will need to present both sides of the argument. This is best done simply by presenting your synopsis of the two arguments in sequence. Thus you might write:

> Derby and Joan (2004) concluded, on the basis of laboratory observations, that the locust jump is a behavioural response triggered by the presence of an attractive mate. However, in a more recent study undertaken in the wild, Bonnie and Clyde (2006) reported that the locusts only jumped to avoid being predated.

In the case of addressing questions still to be answered, you might phrase your statements as:

> Derby and Joan (2004) concluded, on the basis of laboratory observations, that the locust jump is a behavioural response triggered by the presence of an attractive mate. However, it is unclear as to whether this pattern of behaviour would occur in the natural environment.

Your description of the background to the study should, therefore, explain the scientific basis of the study and provide an explanation of what is still not understood and why it is important that it should be understood. On that basis, you can then explain the aims of the study you have undertaken.

It is inevitable that, for many of your practical classes, you will not be aiming to resolve current controversies or discover the solution to a specific question in science, though it is perfectly possible that you may be doing that in your final year research project. For the most part, your practical work will be undertaking experiments that have well-known results. Despite this, it is important that you research the background to the study and present it as a synopsis, in your own words, of the scientific knowledge and can set out the aims of your experiment, based on that knowledge.

Methods

The methods section of the report is where you describe:

- what you did;
- how you did it;
- what you used to do it.

If your report is on a piece of fieldwork, you will also describe:

- where you did it;
- when you did it.

The methods section should be written in such a way that another scientist could repeat your work on the basis of your descriptions. Particularly when writing the methods section for your final year project, you may need to include more detail than would normally be found in scientific papers, for example regarding the composition of solutions used.

Think of this section in terms of preparing a recipe for baking a cake. When you read a recipe, you need to know:

- what ingredients to use;
- how much of each;
- how to mix them together;
- what type of baking tin to use;
- what oven temperature to bake it at and for how long;
- how to analyse the results (eat it!).

Although you are describing what you did, you should still remember to write in the passive voice: 'The locust jump was measured using . . .' rather than 'I measured the locust jump using . . .'.

Check how much you need to write

In terms of reporting on class practical exercises, the necessity for a methods section is variable. For example, where you have been given a detailed schedule to follow, you will quite probably be told that you do not need to copy out the methods again. Even in this case, though, you will be expected to report on any changes made to the protocols, particularly since these may affect the results you obtain.

Assuming that you will be writing a full report, then you will need to describe the methods you used in detail. If you are using a new technique, or have been instructed to give full details of the protocols, then you should set them out in full. In most papers, however, if you are using a standard procedure that has been described in detail in another paper, then you can simply refer to the previous paper and only specify any changes you made:

> The locust leg muscles were stained for ATPase using the method of Peters (1999) but with an extended incubation period of 10 minutes.

Species names

If you are using animals or plants in your experiments, you should give the full name of the species, which is written in italics for the genus and species name, and the number of individuals involved. Where it is significant to the experiment, you should also give other details such as the age, sex, and physical dimensions (e.g. height, weight). Thus for the locust jumping experiment, your methods section might state:

> The jumping experiments were performed using 20 adult, male locusts (*Schistocerca gregaria*), mean weight 1.6 ± 0.3 g (standard deviation).

Human subjects

If you use human subjects in a research project, you need to have ethical approval from a local or NHS ethical committee for the work and you should quote the approval in the methods. You will also need the informed consent of the subjects and this should also be stated.

Chemicals

When listing chemicals or other substances that were used you should give the name of the chemical in full (e.g. sodium bicarbonate) and/or its chemical abbreviation (e.g. $NaHCO_3$), along with the amount used. If you have used specialist reagents you may also need to note additional details such as the name of the supplier.

Specialist equipment

When describing the equipment used, again the amount of detail you need to give depends on the degree of specialization. For example, if you are stating that you used 3 g of sodium bicarbonate, you don't need to give details of the balance used to weigh out the chemical. However, you might have used a specialist piece of equipment for your experiment, for example a confocal microscope or a specific computer interface for data logging. In this case, you should give the details of the equipment such as the manufacturer and the model number. In some cases, it may also be helpful to include a simple diagram of the apparatus used, to show how different pieces of equipment were connected together.

Statistics

We are not going into any detail of statistical analyses as there are many books for the biosciences that address this topic. However, there are some key points you must remember.

For many experiments you will be undertaking several measurements. You will therefore need to give descriptive statistics in your methods. The basic statistics would be measures such as the **mean**, the number of measurements taken (**n**), and the range of the values. For example, if you quote a mean value for a result, you need also to give the number of tests carried out (e.g. $n = 27$) and an indication of the spread of the data. The spread of the data is often given in the form of the **standard deviation**, which gives the reader a measure of the confidence that can be placed in the mean value. In the methods, you would state:

> Means are given with the standard deviation . . . the adult locusts used had a mean leg length of 21 ± 3.2 mm ($n = 20$).

You may also be employing statistical tests in your analysis of the data, for example to distinguish whether a treatment had an effect on the sample population. For most practicals, you are likely to employ routine tests, such as the Student's t-test, or the Mann–Whitney U-test. In such cases it is sufficient to state the name of the test and the level of significance that you are accepting, e.g. $P < 0.05$. For example, in the case of comparison of two groups of locusts you might state that:

> The mean leg length of the adult locusts was significantly greater than that of the juveniles ($t = 3.2$, $P < 0.05$, Student's t-test).

More extensive details regarding statistical analysis can be found in *Research Methods for the Biosciences* (Holmes *et al.*, 2006).

After all that

When you have completed writing the methods section, read through it and check that you could repeat the experiments from the description you have written (think back to the recipe—could you bake the cake?).

Results

The results section is the core of the report. Here, you *describe* and *show* what was found along with the analyses of those findings. You should aim to lead the reader through the findings, highlighting the important features. You do not, however, attempt to interpret or explain the results in any way: this section is a factual description.

When writing the results section, you should use your data to tell a story. It is very important that the text you write can stand on its own in terms of communicating the key points. The graphs, tables, and other images are then used to support the description. A common error in students' results sections is for the data simply to be presented as the graphs and tables with little or no explanatory text, leaving the reader with the task of trying to work out what is important. So rather than write:

> The results for the jumping experiment are shown in Figure 1.

it is much better to write . . .

> From Figure 1 it can be seen that there was no significant correlation between the distance jumped by the adult locusts and their body mass ($r^2 = 0.15$).

In the case of the first extract, it is left up to readers to draw their conclusions from the plot shown in the figure. By comparison, the second extract provides the key summary of the observation, using the figure to illustrate the point and providing statistical confirmation, by means of the coefficient of correlation, that there was no relationship between the two variables.

Calculations and levels of accuracy

When using calculators and spreadsheets, be aware of the levels of accuracy of your measurements: often, a calculation performed on a calculator can result in a number with a string of up to 10 digits. For example, when measuring the jump distance for our locusts we might only be able to measure the distance to the nearest centimetre. So an individual measurement might be recorded as 0.94 m. However, the mean calculated for all the jumps, when using a calculator, might give a figure of 0.942857142 m. If you put this number into your results, it implies that you could record the jumps down to a precision in the nanometre range! As a very simple rule of thumb, don't present calculated values to any more places of decimals than the original measurements.

What is the best way of presenting the data?

This is often a question students find hard to answer. There is a strong temptation to adopt the 'shot-gun' approach: if you include absolutely everything then you can't have forgotten anything! However, for the reader this can make interpretation difficult and someone marking your report may conclude that the reason for putting everything in was because you weren't sure what was important and what wasn't.

Remember that you are telling a story and so try to identify what the key points are, in sequence. Where possible you should present your data in summary form, rather than long lists of raw measurements (which can be included as an Appendix at the end if necessary). If you have measured 20 jumps for an individual locust, the reader doesn't want to see a table of the individual jump lengths. Instead, you can simply present the data in summary form as the mean, the standard deviation, and the number of measurements:

> The mean jump distance for the adult locust was 0.87 ± 0.02 m ($n = 20$).

This statement provides the reader with as much effective information as does the full list of individual measurements and is in a much more digestible form. For a practical report, you may wish to include all the raw data, in which case you can add it as an Appendix to the end of the report.

The three main ways of presenting data are as follows:

- text;
- tables;
- graphs.

Text

Use statements of specific values in the text when you are referring to only one or two items, so the example used above, stating the mean jump distance, is appropriate for presenting in text.

Tables

Tables are very useful for presenting data in an organized form, particularly where you wish to present several variables together to give an overview of a set of results. The layout of the table (e.g. see Table 1 below) is important to ensure that the contents can be easily read and understood. In particular:

- try to avoid having too many columns: having more than five or six columns can make comparison of the data items difficult and also means that the text may be compressed to fit on the page;
- make sure the rows and columns have clear headings, with units, so that the nature of the data is immediately obvious;
- spread the data out and have clear delineation between rows and columns.

Give your table a number so you can refer to it from the text.

TABLE 1 Measurements of the body mass, femur length, and the maximum jump distance for five adult locusts

The title should allow the reader to understand the table on its own

Each column must have a heading to identify the data set and the units of measurement

Locust	Body mass (g)	Femur length (mm)	Maximum jump distance (m)
1	1.34	21.4	0.96
2	1.53	21.8	0.87
3	1.65	21.9	0.88
4	1.72	22.0	0.92
5	1.68	22.0	0.97
Mean	1.58	21.8	0.92
SD	0.15	0.25	0.04

Graphs

There are many different ways of presenting data as graphs. The types of graph you are most likely to use are:

- bar chart;
- histogram;
- scatter graph;
- line graph.

Bar charts can be used to display data related to frequency. So if instead of displaying the relative proportions of each grouping of locust, you wanted to display the actual numbers in each group, you would use a bar chart as shown in Figure 6.4.

FIGURE 6.4 A bar chart showing the numbers of locusts divided approximately in terms of age, based on exoskeleton colour (pink: youngest; brown and yellow: oldest), in the sample tested by the class (*n* = 100).

Axes are clearly labelled to show the data items being displayed

Each column is labelled

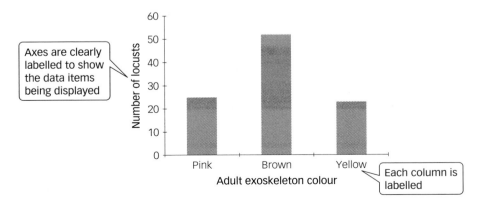

Note that in Figure 6.4 each axis is labelled to identify the information being displayed and that the scale of the *y*-axis (vertical) is set appropriately to display the spread of the data (have a look at the Appendix at the end of this chapter for some examples of how not to draw graphs). A key feature of the bar chart, unlike the histogram (see below), is that the bars are separated by a space. This reflects the fact that the *x*-axis (horizontal) is displaying discrete items and not parts of a continuous population. The horizontal axis is just there as a platform for the bars.

Histograms are also commonly used to display data related to frequency but in this case the variable normally shows a continuous variation. This means that the groups along the *x*-axis form part of a spread of data (e.g. the height or weight of a population), rather than being separate classes, as is the case for the exoskeleton colour. The example shown in Figure 6.5 illustrates the frequency distribution of body mass for the sample of 100 locusts tested in our class experiment. Here, the body masses of the locusts have been grouped into seven groups (or 'bins'). It is important to select the size of the bins carefully so that you can display the data effectively. Avoid having a large number of bins with only one or two items in each. Likewise, having too few bins may mean that the pattern of distribution of the data cannot be seen. You may need to try a few times in rough before you get the best layout.

FIGURE 6.5 A histogram showing the frequency distribution of body masses in the sample of adult locusts tested by the class (*n* = 100). Note that body mass shows a continuous variation and the divisions between the bins are chosen to best display the data.

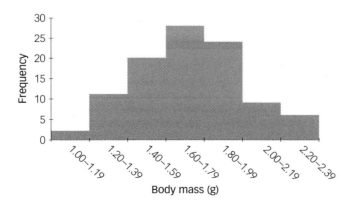

Scatter graphs are often used to display the spread of data for two variables that are related to each other. For our sample of locusts, this might be the relationship between body mass and maximum jump distance, as shown in Figure 6.6.

How do you determine which variable goes on the *x*-axis (the **abscissa**) and which on the *y*-axis (the **ordinate**)? The variable placed on the *x*-axis should be the **independent** variable. This means it is the variable that is already known, or can be controlled by the experimenter. In the locust jump experiment we have a group of locusts of known body mass and we want to see how far they can jump. The body mass is the independent variable and the jump distance is the **dependent** variable, which is placed on the *y*-axis. The dependent variable is therefore the variable that may change as a function of the independent variable. So, in this experiment, we are proposing that the distance a locust can jump depends on its body mass. The logic of this should be clear: in this case, we can assume that the body mass of a locust does not depend on the distance it can jump!

FIGURE 6.6 A scatter graph showing the relationship between body mass and maximum jump distance for adult and juvenile locusts (*n* = 56).

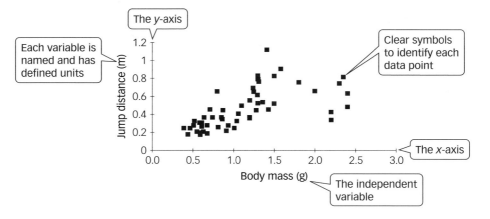

When displaying data as shown in Figure 6.6, there is a further decision to make, which is whether to draw a line to indicate a trend in the data. For example, in Figure 6.6 it appears that, although the data are spread out, there is a trend of longer jump distances being obtained by the larger individuals. Discussion of the processes of line-fitting is beyond the scope of this book: you should refer to the companion text on statistics (*Research Methods for the Biosciences* by Holmes *et al.*, 2006). As a rule of thumb, however, be wary of the complex line-fitting programs that can be found in many spreadsheet and statistical software packages. These programs can fit scattered data with very complex mathematical equations, which can be very difficult to interpret. If you think there may be a simple trend such as a linear relationship, then use the program to calculate the linear regression line that best fits the data. However, you should only draw conclusions regarding the linearity of the relationship if you also calculate the coefficient of correlation and this demonstrates a statistical significance.

Line graphs. Rather than having a scatter of data points, as for the population above, your experiment may have involved taking a series of measurements against a set baseline. An example of this would be taking measurements at set time intervals. For such data, time is represented on the *x*-axis and the change in the variable on the *y*-axis. In the case of the data shown in Figure 6.7, the experimenters have measured the power generated by the leg extensor muscles as a function of joint angle. The measures have been repeated for 10 individuals and then averaged, so each data point actually represents the mean of the 10 individual results. This plot therefore illustrates another important feature of graph-drawing which is that, for averaged data, you should also indicate the spread of the data items contributing to each value displayed. This is done by drawing in the standard deviation (see *Statistics* in the *Methods* section, above) as vertical error bars above and below the data point. The reader can then see how reliable the plotted points are in illustrating the trends being shown.

FIGURE 6.7 A line graph illustrating the relationship between the mean power output of the locust extensor muscle and the initial joint angle for 10 adult locusts. The error bars show the standard deviations of the means. There was a significant negative correlation between power output and joint angle ($r^2 = 0.91$, $P < 0.01$; method of least squares), as indicated by the linear regression line.

As for the scatter plot, you will need to decide whether to draw a line to illustrate a trend in the data. This can be done either by simply joining the individual points together, or by calculating the regression line that shows the best fit to the data, as has been done in Figure 6.7. Again, you should only plot a best-fit line if there is a statistically significant coefficient of correlation. This coefficient value should be given in the figure legend.

Checklist for graph drawing. Make sure that:

- you have used the appropriate form of graph for displaying your data;
- the data are displayed on the correct axes—the independent variable is on the *x*-axis, the dependent variable on the *y*-axis;
- the axis scales are appropriate for the spread of the data;
- the data points and the axis labels are large enough to be clearly legible;
- the axes are labelled and the units of measurement given;
- the graph has a clear figure legend or title so the reader can appreciate what is being displayed without needing to read the accompanying text.

Look at the Appendix at the end of this chapter (p. 107) for some examples of how not to draw graphs.

Using images

You might use a variety of images, such as photographs, in your reports. These photographs may be taken using microscopes or electron microscopes or they may be photographs of a landscape. The key rules here are to make sure that:

- the illustration has a legend or title that explains what is being shown and how the image was obtained;
- any specific aspects of interest are identified by arrows, so the reader can appreciate what is being described;
- there is an indication of the scale of the image—this is usually done by adding a labelled scale bar to the illustration.

Results checklist

After you have completed the writing of your results section, read it through carefully and ask yourself the following questions:

✔ Have I clearly described the key findings in the text?

✔ Do the tables and graphs present the data in a clear form?

✔ Are the key conclusions supported by appropriate statistical tests?

✔ Have I avoided trying to interpret the findings?

Discussion

The discussion is the section of the report where you interpret the findings from your experiment and place them in the context of the research literature. When it comes to writing up a practical class or project, this is often the most demanding part of the exercise because you must display your understanding of the work you have done and how it fits into the studies undertaken by other scientists.

A useful starting point for the discussion is to write an introductory paragraph that summarizes the key findings from the results:

> In this investigation into the mechanisms of the locust jump, it has been shown that . . .

This paragraph then provides a good link with the results section and gives you a clear list of the points you need to interpret. There should also be a link back to the aims that you set out in the introduction.

Having identified your key findings, you then need to discuss each of them in turn. As for the introduction, you will need to refer to the relevant research papers, highlighting their findings and relating them to what you have found. It may be that your findings are different from, or even contradict, those presented in other papers. In such cases you will need to explain why you think this may be so: don't always fall back on the staple explanation that you probably got something wrong! Whereas that may be true in many cases, there may also be genuine reasons for your experiments having generated different results. If you are writing up a research project, it is likely that you will have new data and these will need careful interpretation with reference to the research literature.

You should remember to round off the discussion with a concluding paragraph that summarizes both the key findings and their interpretation. In class reports, you may also be encouraged to identify what experiments you might do next in order to progress the work. An example of a short conclusion for our locust jump experiment might read as follows:

> In these experiments, it has been demonstrated that there is no significant correlation between the body mass of adult locusts and the maximum distances that they can jump. These findings have been discussed in relation to the studies by Smith and Jones (2004) who reported that the processes of power generation by the adult locust could be more closely related to age rather than body mass. The relationship between . . .
>
> There was a large range of variation in the results obtained in these experiments and any future study should employ more locusts, with more jump trials for each locust, in order to increase the probability of demonstrating significant differences between the different groups. Furthermore, the failure of the heating system during the course of the experiment meant that the later jumps were being performed at temperatures that were sub-optimal for the locusts.

References

As with all your written work, you must include a full list of the references that you cited in the text (see section 3.4, *Citations and references*).

6.3.3 **Other sections**

You may wish to include some additional sections to the report. The most common of these would be an Appendix and a set of Acknowledgements.

Although the results section is where you present your data, this is often in the form of processed information: for example, the means of sets of measurements. However, you may wish to include all the raw data, to demonstrate the full set which you have obtained and from which you have derived the summary results. You may also wish to include worked examples of your calculations, such as the statistical tests. Under such circumstances, these can be presented in tables in an *appendix*. This is useful in a practical report but would not be typical practice in a research paper.

It is unlikely that you will need to include an *acknowledgements* section in a report from a class practical, but this is more common in a project report. For your project you may have been given specific support by academic or technical staff or have been given access to specific items of equipment that would not normally be available to you. It is then a matter of good practice to acknowledge this help in a brief statement at the end of the report. This should not be treated like the eulogies at the Oscar ceremonies: just give brief, factual acknowledgement of any specific support you were given.

6.3.4 **Write the sections in an appropriate order**

Writing a report can often seem a daunting process, particularly if it is for a large piece of project work. Before you put pen to paper (or fingers to keyboard), check again any brief you have been given for the structure of the report so you know exactly what is expected of you.

Some sections of the report should be much easier to write than others. For example, the methods section is simply a factual description of what you did and so this is often a good section to start with.

The next section to tackle is the results. You should think carefully about how you want to present the data first of all and then produce your graphs and tables and carry out any statistical analyses. When you have the data presentation sorted out, the process of writing is fairly straightforward since you are aiming to describe the results that you have in front of you. Don't forget that you are telling a story in which the storyline is illustrated by the figures and tables.

Having set out the key findings of the experiments in the results section, you are then in a position to consider their interpretation. Before beginning to write the discussion, though, you should read through the relevant research papers so that you have a clear picture of the relationship of what you have done to the current research. Take each of your key findings in turn and write about it, making sure you explain the relationship with the previous research in order to interpret what you have found.

The final sections to write are the introduction and the abstract. Having written the discussion, you should be in a position to present the background to the work and to set out the aims. As for the discussion, you will need to refer to the relevant research articles to support your statements. This should lead logically into the aims of the experiment and you can check off the aims against your final conclusions. The abstract is the final piece to be written, based on a very brief synopsis of the other sections of the report and highlighting the key findings of the research.

Finally, as with all written work, it is always a good idea to put the completed report to one side for a few days and then to re-read it, checking that it all ties together and could be understood by someone who had not done the experiments.

✳ Chapter summary

In this chapter we have highlighted the importance of preparation before a practical class and what to do during a practical class to make the write-up afterwards an easier process. When writing up, it is important to use an appropriate writing style and to adhere to the conventions of scientific report structure. When writing the report it is not necessary to write it in order—in fact there are sections (the abstract being the most obvious example) that can only be written last, and there are other sections that come in the middle (the methods and the results) that are easier to write first. Finally, read through and check your report to make sure that it tells a logical story and that there are no errors or omissions.

Appendix: examples of how not to draw graphs

The display of data in graphical form is a key part of the preparation of many reports of practical work, whether it be for enzyme assays or the populations of birds in a cliff nesting site. In the section on graphs, we have described some of the key principles in drawing graphs but there are many pitfalls, especially when using spreadsheets or graph-drawing packages. Some of these pitfalls are more obvious than others. Have a look at the graphs shown below as Figures 6.8 and 6.9 and list the errors that you see: each shows at least one error!

FIGURE 6.8 Power output against joint angle, locust experiments.

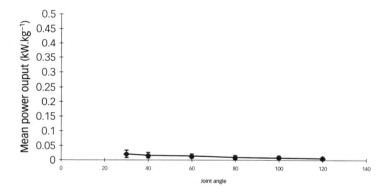

FIGURE 6.9 Scatter plot showing the relationship between body mass and maximum jump distance for adult and juvenile locusts, *Schistocerca gregaria* (*n* = 56).

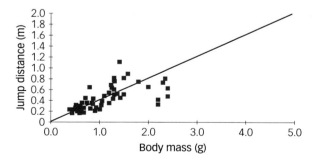

Figure 6.8 shows a number of common errors (though rather exaggerated) that might be made when displaying the data shown in Figure 6.7.

- The figure legend is too brief and does not explain what is being illustrated. The test here is: if the reader only had the figure and the legend to go on and no accompanying text, would he or she be able to understand what was being shown?

- The scale for the *y*-axis is inappropriate. The vertical spread of the data points does not even extend to the first scale point on the axis. Furthermore, the compression means that the error bars are not actually visible. The scales should be set so that the data points extend over almost all of the length of the axes.

- The font size for the scale on the *x*-axis is too small to be legible. Make sure that the fonts for the two axes are legible and are the same size.

- The *x*-axis label 'Joint angle' does not specify the units.

- The numbers on the *y*-axis are not set to the same level of precision. Make sure that the same number of decimal places is specified for the scale indicators.

Figure 6.9 at first sight might appear to be okay since the axes are appropriate and clearly labelled and the legend informs the reader as to what data are being presented. However, there are some key flaws in the presentation:

- A straight line has been drawn through the data points but there is no indication as to the basis for drawing such a line: was it simply drawn by eye as appearing to be the line that best fitted the data, or was it calculated using linear regression? The method for arriving at drawing such a line needs to be stated in the figure legend. If it is drawn by linear regression, then the method chosen for calculating the line needs to be stated and the coefficient of correlation given so that the reader knows whether the linearity is statistically valid. An example of the statement to include in the figure legend would be as follows:

Statement of linearity

Scatter plot showing the linear relationship between body mass and maximum jump distance for adult and juvenile locusts, *Schistocerca gregaria* ($n = 56$, $y = 0.4x + 0.02$, $r = 0.62$, method of least squares).

Linear equation for the plotted line

Coefficient of correlation

Method used for the regression

- Is it valid to extrapolate or interpolate a line? The straight line here has been drawn back to the origin (coordinate 0,0); this is **interpolation** and it makes the assumption that the linearity of the relationship continues below the range of the data set on which the calculation was based. You therefore need to ask yourself whether this assumption is justified or in this case biologically possible—after all, how many locusts are there that weigh 0.01 g? The same argument holds true for **extrapolation**, the process of extending the theoretical relationship beyond the range of the data set. Do you know of any locusts that weigh in at 5 g? Apart from the fact that they would be big and scary, can you be sure that they could jump 2 m? The rule of thumb is that if you draw a straight line through the data set, it should only extend over the range of the data points and not beyond unless you have strong evidence that the relationship will hold true for other data points.

Chapter 7

Working in tutorials and groups

➔ Introduction

Tutorials and group work form key elements of teaching within most degree programmes and can be employed in a variety of formats, as we will see in this chapter. Irrespective of the specific format of a session, the most important things you can do are to **prepare** and to **participate**. Without preparation and if you are not willing to participate, you will derive little benefit from a tutorial or group work session.

Tutorials can be quite a daunting prospect as many students are unsure of their knowledge base and ability to discuss ideas, particularly in front of a member of staff who may be a world-expert on the subject. But you should remember that your fellow students will be in the same position and your tutor will want to encourage you to participate.

Tutorials and group work are useful in developing a range of important skills that will be of value to you in your studies and also in your career development. These skills include the development of:

- subject awareness—increasing your knowledge of the subject;
- critical thinking—listening to different ideas as they are put forward, thinking about them, and evaluating them in terms of the strengths and weaknesses of the arguments;
- communication skills—presenting your own ideas in a way that can be understood by the rest of the group, and learning the essential ability to be able to refine the views expressed by other people, pick out the key points, or even counter them, without causing offence.

We will begin by addressing tutorials.

7.1 Tutorials

The basic format for a tutorial is a relatively informal group where a tutor meets with a small number of students to discuss specific aspects of the course or module being studied.

7.1.1 Different types of tutorial

There are two main types of tutorial.

- Academic tutorials that are normally linked to the modules you may be taking at the time. These sorts of tutorials may very occasionally be run on a one-to-one basis but will more commonly involve five or six students or even up to 20 or so. Academic tutorials may involve a range of activities such as discussions about a specific topic, short presentations given by members of the group followed by a discussion, or the consideration of the answers to previously circulated questions.

- Tutorials held with a personal tutor, or equivalent, to discuss your progress on the course. These will often be on a one-to-one basis and may include discussion of career issues and personal development planning, as well as aspects of the non-academic areas of university life, such as accommodation.

As well as tutorials, you are very likely to be required to participate in various forms of group work. Many of the principles, particularly in terms of engagement, are the same for the two types of session, but we will also consider group work separately later in the chapter (section 7.2).

7.1.2 Preparing for tutorials

Preparation for tutorials is essential. It may well be that you have been given a specific topic to research, a worksheet to go through, or some sample questions to answer. On the other hand, you may simply have been told what the theme of the tutorial will be. Either way, you need to be prepared, so that you can benefit from the discussions and participate in a meaningful way.

Preparing for academic tutorials

Academic tutorials offer you the opportunity of discussing your subject with a bioscientist, who is very knowledgeable, as well as with your fellow students, in a fairly informal setting. If you are going to be able to engage in that discussion and to learn from it, then careful preparation is essential. It is likely that you will be given one or more specific topics that will form the starting point for the tutorial discussion. As with the preparation for writing an essay (see Chapter 5, *Writing essays*), or giving an oral presentation (see Chapter 9, *Preparing scientific presentations*), you need to read up on the topic and make succinct notes that you can use as prompts in the discussion.

If the tutorial topic is based on material that you have been covering in your lectures, then the obvious place for starting your reading will be your lecture notes, before moving on to read any specific references you may have been given. Have a look at the sections on *Using textbooks*, *Reading a research paper*, and *Note-making strategies* that are covered in Chapter 3, *Working with different information sources.* As a guide, the most important things to ensure are that:

- you are keeping focused on the question asked;
- you are making brief notes that summarize the key points and which you will be able to use as prompts in the discussion;
- you are keeping a record of the sources of the information (see the section on *Citations and references* in Chapter 3) so that you can cite the paper or text if you are challenged regarding the reliability of the information you are presenting, or if you want to find the information again to get more detail.

A short time before the tutorial read through your notes to make sure that you understand the points and to refresh your memory.

Preparing for personal tutorials

Personal tutorials may cover a range of topics. In relation to the themes of this book the most important aspects are likely to be topics such as your academic progress, module choices, personal development planning, and thoughts on careers. Again, if you are going to benefit from the opportunity then you must prepare in advance and then be ready to discuss these issues openly.

For example, in order to be able to discuss proposed module choices for a forthcoming semester or academic year you need to have read about the range of choices that are open to you: this will include checking whether taking the modules require you to have taken specific pre-requisite modules in the current year of study. You also need to have thought carefully about where your strengths and interests lie and where you think studying those modules might take you in career terms. In the context of your development, you should also be thinking about whether you would benefit from further training to develop your skills: for example, if the modules are assessed by poster presentations and you are not sure how to go about preparing posters, then you should read Chapter 11, *Creating academic posters*, in this book and also consider seeking guidance from your tutors or perhaps from the university's learning support centre, or equivalent.

7.1.3 **During tutorials**

You have come to the tutorial, having done your preparation. The next important element is to make sure that you participate. Participation, of course, means making active contributions to the discussions, but it also means listening carefully, thinking about what is being said, and relating that to what you have already learnt.

Contributing to academic tutorials

The first and most important thing to remember about tutorials is that they are not mini-lectures: your tutor will expect to have a role in facilitating and guiding the discussion but will not expect to be talking for much of the time. Indeed, some tutors are experts at sitting in silence waiting for the students in the group to start talking! There will, therefore, be an expectation that you will be an active participant—that you will attend the tutorial having prepared beforehand and be prepared to discuss your ideas, as well as to listen to the points made by your peers and the tutor.

You will probably find that this is all quite a daunting prospect at first and there is a strong temptation to sit looking at the floor, hoping not to be noticed! While this is an understandable response, it is not a very productive approach. Remember that your tutors are there to help you understand the subject and will want to help you develop your skills in communicating ideas; however, they can only do that if you participate. Don't be afraid of expressing ignorance if there is something that you don't understand: the only thing that is likely to be irritating to the staff is if it is apparent that you have not bothered to do any reading or preparation beforehand.

Depending on the structure of the tutorial, there may be some short presentations about topics that you and your peers have been asked to prepare in advance, followed by discussions. Alternatively, there may be specific questions that you have been asked to discuss, so the whole tutorial takes the form of a facilitated discussion. Irrespective of the format, there are key ways of engaging with the process.

- Listen carefully to any ideas being presented by the tutor or other students and note down the key points of the arguments.

- At the end of the statement, say whether you agree or disagree with the ideas, presenting evidence to support your views. It is no good just saying you agree or disagree with an idea: you also have to be able to justify that view.

- If there are things you don't understand, then ask for further details or explanation: never leave the tutorial not understanding things because you were afraid to ask questions.

- At the end of the discussion, make sure you can summarize the key points for yourself.

- You do need to be sensitive, though: if you are a confident speaker, don't try to dominate the discussion but let other people express their views as well since that is an important way of learning.

Contributing to personal tutorials

With personal tutorials, as with academic tutorials, you can only expect to benefit if you engage actively in the process and, in these cases, are prepared to be open in terms of your views and aspirations, as well as being realistic in terms of your abilities. This means that you may need to open up discussion topics and explain your background to them, rather than waiting for your tutor to raise them: despite their best intentions, they may not be able to guess what is on your mind or worrying you at the moment. You should also recognize that your tutor may not have all the answers. Even if they don't have the answers, however, they should be in a position to be able to advise you on where to seek further guidance.

It is important to remember that the personal tutor is there to try to help you. Often students are anxious that if they raise issues of concern with their personal tutor, it will affect how the department views them. This should not be the case under any circumstance. Although the personal tutor is not bound by a confidentiality clause as are, for example, GPs or counsellors, they have a duty to be discreet about things they are told and not to be judgemental.

7.2 Group work

The process of collaborative or group working may often be very productive. Indeed, there is a wealth of evidence that material learnt through group work is retained better than that learnt through formal teaching methods. Group work is often very useful in developing critical skills because as a member of a group you need to evaluate the contributions that other members of the group have made, as well as reflecting on your own contributions and discussing their validity with the rest of the group. In the biosciences, group work is very common as a part of laboratory classes where you typically may have to work in pairs or threes in order to complete the experiment. In the case of group work based around discussion questions, many of the points relating to tutorials may also be applied. In particular, you need again to be thorough in your preparation and be prepared to engage actively in discussions.

One of the additional key features, though, relates to organization and the dynamics of interaction of the group. Unlike the tutorial, where your discussions are facilitated by a tutor, group work exercises will normally require you to work with your group on a set task, either in a classroom setting, for a relatively short time, or as an extended exercise which you undertake at times and locations of the group's choosing. The dynamics of how people interact and contribute to the group are a very important element contributing to the overall success of the group. Under these conditions, how you behave in the group is also clearly important. Some key points to remember here are:

- keep an open mind and be prepared to listen to, and consider, the ideas put forward by others in the group;
- avoid being judgemental—for example, by being dismissive of someone else's ideas;
- be careful of your body language—for example, folding your arms and staring out of the window does not convey a positive message.

Let's now consider each of these types of group activity.

7.2.1 Short-term group work

Short classroom exercises are often employed as a way of breaking up lectures or as a means of checking that students have understood a specific concept. You will probably be asked to talk to the people around you to discuss a specific topic or answer some questions. Many of the issues of group dynamics and team-playing that are discussed in the next section are not applicable here. For the most part, the most important skill is remaining focused on the topic and using the time constructively: it is very easy for discussions to degenerate into a review of the latest gossip and, although that might be of great interest, it will not help get the task completed! It is also often very easy to spend a long time debating the first one or two aspects of the question and then find you have run out of time before completing the exercise, particularly if you only have a few minutes for your discussion.

There are several ways of providing some structure to your discussions and making sure that you get some answers whilst also getting everyone involved as far as possible.

- If you have been given several questions to answer, it is very helpful to agree quickly at the start how long you are going to spend on each question, and make sure that you adhere to that time-scale, noting down your agreed answers as you go.

- A useful way of getting ideas quickly and of engaging everyone in the discussion is to go round the group in turn with everyone putting in an idea or point of information related to the question.

- An alternative is to brainstorm: people in the group mention ideas in any order as they think of them. This can be very productive if some creative thinking is needed, because the lack of formality encourages more open thinking. It can also generate numerous ideas very quickly. However, it may also lead to some people feeling excluded because, for example, they feel apprehensive about the merit of their ideas or because they are too shy to contribute. In such situations, try to be aware of the other members of the group and encourage non-contributors to participate: after all, they might have the best ideas of everyone! As with the guidance for behaving in tutorials, the key points here are to listen to what others have to say, to ensure your body language is appropriate and open, and not to be dismissive even if you disagree with what the other person is saying.

7.2.2 **Longer-term group work**

If you have been allocated a task that will require you to work in a group over a period of days or weeks in order to complete the task, then the dynamics of the group become a very important factor contributing to the success, or otherwise, of the exercise. Many students are dubious about the benefits of group working, particularly if the exercise is worth a significant amount of marks. However, the skills involved in such exercises are very important in career development because team-working skills are among the most highly valued by employers.

Undertaking a specific project as a group can be broken down into a series of key stages, as shown in Figure 7.1. We will consider each stage in turn.

Get off to a good start

It may be that you have been allowed to select who you work with, and have chosen to work with a group of friends. On the other hand, you may have been put into groups

FIGURE 7.1 Stages in the process of group work for a specific project.

by your lecturer and you may or may not know the people you are having to work with (and, if you do know them, you may not like them!). You might not feel happy if you find yourself grouped with people you don't know, or don't like, but it can be very good experience for your working life, since you may well have to team up with strangers to undertake a task in your employment, or you may have to work with people you don't necessarily like. In either case, the work still has to be done.

Time spent in ensuring that the group gets off to a good start should pay dividends as the project develops. Setting up formal structures for the group—like having a group leader, agreeing task allocations, or having meeting agendas—might seem rather over-the-top for some projects, but it can all be useful experience. Additionally, having formal aspects to the project, which are agreed in advance, helps to avoid problems later on.

Get to know each other

This first stage is clearly most important if you are going to be working with people you don't know. Even in a group where you know everyone, that might not be the case for everyone else. So, it is very helpful to get together for your first meeting, away from the work environment, perhaps in a coffee shop or pub or just sitting on the grass if the weather is good! Use that meeting to ensure that everyone feels part of the group and can, at least, begin to feel confident about working together. If you need to do a round of introductions, get each person to say something about themselves as a 30 second biography or say what they think might be particularly interesting about the project.

Establish your ground rules

Don't be misled into thinking that, just because you are working with friends, you can all just get on with the task. Whether you are with friends or strangers, it is essential that you agree the ground rules for running the project, otherwise things can come unstuck very quickly. The extent of these rules will depend on how long the project will be running for. If it is several weeks, however, you might like to consider the following questions/points which could be set out in a generic agenda for the first meeting (as shown in Figure 7.2).

- Do we need a project leader/coordinator who will keep track of progress?
- Would someone take the role of secretary to keep a record of points agreed at meetings (as shown in Figure 7.3), and make sure meetings stick to time (it doesn't have to be the same person each time)?
- Do we need formal agendas for future meetings? If you think that setting agendas makes the whole process seem too formal, then you could have meeting plans.
- What time commitment should each member of the team be making to the project?
- What action should be taken by the group if one or more members are not contributing?

FIGURE 7.2 Proposed plan for the group's first meeting.

BioStars Project Group Meeting Plan
14:30 Tuesday 2 May
Biology dept coffee bar

1. Introductions
2. Group organization
   a) Do we need a group leader?
   b) Do we need a secretary?
   c) Do we need . . .
3. Group operation
4. Brainstorm
5. Division of project tasks
6. Key progress dates
7. Next meeting

FIGURE 7.3 Example of first section of notes from the group's first meeting.

BioStars Project Group 1st Meeting Report
14:30 Tuesday 2 May
Biology dept coffee bar

1. Present
Alex, Komal, Sarah, Lee, Rathan

2. Organization
Komal will act as group leader, Sarah will take notes

3. Operation
Project task is to present a critical review of deep brain stimulation in the treatment of Parkinson's disease. 6 weeks working time.
Group to meet every week to update

4. Brainstorm
PD – degeneration of dopaminergic neurons in substantia nigra (basal ganglia) movement deficits . . .

Brainstorm the project

You have been given the project title and a brief for what is expected of you. No doubt different members of the group have different ideas about what the approach should be, what needs to be done, and how. Use a brainstorming session to let everyone have their say. It is important that the discussions at this stage are free ranging, so that the full breadth of the topic can be covered and also that everyone feels free to participate. So, at this stage, all contributions should be treated equally and there should not be any criticism of the ideas that are offered: you will have time later on to focus on specific aspects.

(Note: it is important that someone records the ideas, otherwise all that brain power will be going to waste! However, you need to make sure that the person recording ideas has the chance to contribute ideas as well and is not just there as a note-taker. Also, set a time limit for the discussion, otherwise you could be there a long time!)

A useful approach to brainstorming is to give everyone 2 or 3 minutes to think about the problem and then go round the room so each person puts their ideas in. These ideas can be recorded as bullet points on a flip chart which everyone can see. When the first round of ideas is up, see if there are links between them so that you can group them together. From that grouping, you can then go round the room again to build on these with further ideas: but remember your time limit!

FIGURE 7.4 Identification of the tasks involved in preparing the presentation.

> **Tasks for Parkinson's disease project**
>
> **Research topics**
>
> 1. Structure & function of the basal ganglia
> 2. Neurological features of Parkinson's disease (PD)
> 3. Current drug treatments
> 4. What is deep brain stimulation? (DBS)
> 5. Use of DBS in treating PD
>
> Identify the key points for the presentation
> Prepare the presentation and write the text
> Prepare handouts for the audience
> Rehearse and review the presentation

Agree tasks

Once you have the outline of how to approach the topic, you need to break this down into a series of tasks, so that you can plan the way the work will be done and who will do it (as shown in Figure 7.4).

One of the obvious features of group projects is the division of labour so that there is co-operation; if you have everyone trying to do everything, there is no real benefit from being in a group. Again, it is vital that, right from the outset, the group agrees who is going to do what and that a common record is kept of the division of labour (Figure 7.5). So, having drawn up a list of tasks from your brainstorming session, the group needs to agree how they are divided up so as to be reasonably equitable. At this stage it is really helpful to take account of different people's particular skills: if you have a successor to David Attenborough as a presenter, then it makes sense for that person to be taking a major part in any presentations.

Plan your timeline

As well as dividing the project up into different elements that individuals will be responsible for, you also need to divide it up into chunks of time so that you have specific milestones that you can identify in order to record progress. It may be helpful to link this in with the agreed division of labour, so that each person has a target to work towards and take responsibility for (Figure 7.5). Also, build in some slack in the timing, so that you allow some extra time at the end of the project for completion. For example, you could aim to complete the project a week before the deadline, then if there are any problems you do have some leeway.

FIGURE 7.5 Key dates for the project and breakdown of project tasks.

Project timeline:

May 9th	Completed background research
May 16th	Agree the bullet points for the presentation
June 6th	Draft presentation complete
June 10th	Rehearse presentation
June 13th	Class presentation!!!

Work division:

May 9th

Komal – neurological aspects of PD
Alex – current drug treatments
Rathan – what is deep brain stimulation?

. . .

Keeping going

It is often the case that everyone is pretty enthusiastic about the project to begin with and the first meeting goes well with agreed plans of action and so on. Keeping the momentum going, however, can be much harder. This is especially true when more imminent coursework deadlines focus input elsewhere and the long-term nature of the group project tends to give rise to an element of complacency. This is when it becomes important to monitor—and stick to—your original project plan.

Meet regularly

Keep to your schedule of regular meetings. In particular, don't let a meeting be dropped because people have not done their work (even if one of those people is you!): it is very easy to let this happen, and you will quickly find you are on a slippery slope and no progress is being made. If nothing else, therefore, the meetings act as a reminder that the work needs to be done and that the group needs to keep on schedule otherwise the project will be compromised.

Keep a check on progress

At each meeting, it is important to review exactly where the project is and whether you are on track. Refer back to your timeline (Figure 7.5) and check off your progress against the target dates.

This checking, however, is not just a tick-box approach, blindly accepting that the work has been done. If the group is aiming for high quality in its performance, then there also needs to be an evaluation of the quality of the work each person has produced. If your tutors gave you a marking scheme or equivalent, then the members of the group need to evaluate each section of the work that has been produced against the marking scheme. Though it may seem like it, this is not just a chore: it is important for your own learning that you have an understanding of the whole project, not just the sections you are responsible for. It is also a very valuable learning exercise in developing the skills of critical evaluation, reflection, and discussion.

People who fall behind

One of the biggest, and most common, problems with group work is dealing with group members who don't keep up with the work, or fail to engage with the group. There are various strategies the group can adopt to try to resolve the situation.

- Talk to the person. This is the first and most vital approach. If the person comes to the meeting, then you can discuss the situation. If he or she has stopped attending the meetings, then you need to try to re-establish contact as soon as possible.

- Find out what the problem is. It may be that the person has been ill, or struggling with something else and can catch up, particularly if given a bit of time or some help as a one-off. If there is no real excuse, then the group needs to refer back to the ground rules and make it clear that everyone needs to fulfil their agreed contributions, otherwise the whole project may fail.

- Persistent non-co-operation. If one member of the group will not co-operate, even after you have discussed the position, or if that person refuses to respond, then you need to take more explicit action. Probably the best option here is for you to contact the tutor and explain the situation and ask for assistance in resolving the problem, though this may be a move of last resort as it indicates that the group can no longer sort out its own problems.

The final product

The deadline is approaching and you now need to bring everything together ready for submission. If you have let things slip, then the quality of the final product may be lacking. If, however, you have kept to your timeline, this phase of the project should allow you to produce a really polished piece of work.

- Bring everything together: one of the difficulties of group work is that everyone has their own style of writing or expressing themselves but the final piece of work needs to appear as a consistent piece and not as three or four completely separate elements that have been lumped together. This may require you to edit the sections so that the approach or style of writing is the same and also so that there are sensible links between the separate sections. Editing by committee is often a lengthy and unsatisfactory process. In most cases, it is preferable for one person to be responsible for bringing the sections together and editing them to create a single 'voice': think back to the initial process of agreeing the tasks so that you could play to the strengths of each member of the team.

- Rehearse your presentation (see the section on *Practice* in Chapter 9, *Preparing scientific presentations*): if your final product is a presentation, then it is essential that you rehearse. In particular make sure that:
  - the presentation keeps to time;
  - the illustrations all show up appropriately. This is especially important if you are using animations or, for example, showing video clips that are embedded in the talk;
  - if you have more than one speaker, make sure that the links between them are smooth and everyone knows who is speaking in turn.
- Check against the brief: before you submit the work, run through the instructions you were given at the outset again, and make sure that you have addressed all the criteria. It is important that, as far as possible, everyone agrees that they are happy with the final product and are prepared to sign it off.

When it's all over don't forget to make use of the feedback (see Chapter 12, *Using feedback*) and think about where any problems were and how you might do it better next time.

....

✳ Chapter summary

In this chapter, we have looked at ways of working for tutorials and group work. The key message for all these activities is that you need to prepare, to participate, and to listen to others.

Preparing for academic tutorials means reading through any lecture notes or reference material in order to make sure that you know the subject well beforehand. For both academic and personal tutorials, you also need to think about what the objective of the tutorial is. During the tutorials, make sure you listen to what others have to say and be prepared to take their ideas on board as well as argue against them if necessary. Also, remember to take notes, so that you have a record of the discussions.

For group work you must be prepared to work with other people, even if you don't like them! Agree the tasks to be done, who will do them (work to each person's strengths) and by when, so that the group can keep a track of how the project is developing. In order to do this, you must keep meeting regularly and checking progress against your timeline: this is particularly important if the project is running over quite a long period.

Chapter 8

Avoiding plagiarism

➔ Introduction

This chapter addresses the increasingly common problem of plagiarism. We will begin by defining what plagiarism is and then look at some of the reasons students give for committing plagiarism in their work. We will then outline a method for you to avoid plagiarism and highlight some of the many good reasons there are for making sure you do. It is important to note that, although most forms of plagiarism probably occur in the context of essays and practical reports, it is possible to plagiarize in any form of communication; so the study skills required to help you to avoid plagiarism in your writing are essential to your success in every area of your academic work.

8.1 What plagiarism is

Plagiarism is an increasingly common problem in universities. The increase is partly due to the internet and partly due to the copying and pasting features of word-processing applications, making copying from one source to another a very easy process. In addition to this, the academic conventions surrounding plagiarism are less widely understood; students often arrive at university with only a very limited understanding of the issue. Because of the rise in the incidences of plagiarism, universities are much more aware of the matter and are therefore keen to address it. Many universities are now using plagiarism detection software to detect cases, which has made it much more difficult for students to get away with plagiarism. However, it is important not to get the issue out of perspective; as we shall see, plagiarism is often not deliberate and there are many positive reasons for learning how to avoid it. But first, we need to define what the term 'plagiarism' means.

8.1.1 Definition

The *Oxford English Dictionary* defines plagiarism as:

> The action or practice of taking someone else's work, idea, etc., and passing it off as one's own; literary theft.

So plagiarism is about using someone else's work but giving the impression that it is your own. Let's think for a moment about the different elements of this definition:

- *taking*—could be from a book, journal, internet site, or even a lecture or handout;
- *someone else's work*—could be someone else's words but could also be their ideas, data, or images;
- *passing it off as one's own*—could be either deliberate or accidental and could be in the context of an essay, a lab report, or a presentation (in fact, any assessed piece of work);
- *literary theft*—tells you in unambiguous terms how it is perceived by many, including your tutors.

So put simply, plagiarism is a form of cheating. You wouldn't cheat in exams and you shouldn't cheat in coursework either. Therefore, just as there are penalties for cheating in exams, there are penalties for cheating in coursework too.

8.1.2 Academic method

Why are universities generally, and academics in particular, so concerned about plagiarism? Why is it considered to be such a big deal? Are they over-reacting or are there good reasons for their concern? Well, let's think for a minute about what underpins academic work. To be more specific, let's consider what is sometimes known as the 'academic method'.

It was Isaac Newton who said of himself (in a letter to Robert Hooke in February 1676):

> If I have seen further it is by standing on the shoulders of giants.

By 'have seen' he was referring to the tremendous discoveries he had made during his lifetime; by 'standing on the shoulders of giants' he was largely referring to the people who had gone before him whose ideas he developed and built upon. In making this statement Newton was highlighting a very important principle of academic study: that although you might research and then write-up something individually, you are in fact contributing to a collective endeavour. Even the most novel or original of discoveries will build upon the ideas of others, and it is important that those ideas are acknowledged and that it is only your own contribution that you claim for yourself.

So, plagiarism is an important issue because academic honesty underpins the academic method: it is fundamental to all academic work. If you think about it, this also means that if you can learn the study skills that enable you not to plagiarize, you will be helping yourself not only to avoid being penalized, but, more positively, to perform well in all areas of your academic work.

8.1.3 Deliberate plagiarism

When people hear the word 'plagiarism', they usually think of someone deliberately copying text without referencing it, but, as we have seen from the definition above, plagiarism is often much broader than simply copying text. We will deal first with examples of deliberate plagiarism.

Copying

Imagine that a student used the following text in an essay on 'Factors influencing the development of cancer'.

> There are many factors which influence the development of cancer. These include both endogenous factors such as inherited predisposition and exogenous factors such as exposure to environmental carcinogens and infectious agents. Another factor which has a clear influence on the type of cancer which develops is age.

The first sentence is the student's own but the next two sentences are not written by the student; rather they are taken from the *Introduction to Cellular and Molecular Biology of Cancer* by M.A. Knowles and P.J. Selby (2005). The student has therefore copied the text without any indication that some of the words have come from another source. They have therefore plagiarized by copying.

Alternatively, take an example of an image: imagine a student had used Figure 8.1 in an essay on 'Factors influencing the development of cancer'. If the student failed to acknowledge the source of the image, in this case Knowles and Selby (2005), Figure 1.3, p. 8,

FIGURE 8.1 Sample of a plagiarized image.

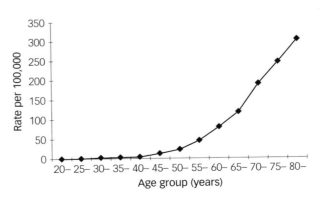

they would also be plagiarizing by copying. It is important to remember that plagiarizing by copying images is equivalent to, and so just as serious as, plagiarizing by copying of text.

Collaborating

It is also possible to plagiarize by collaborating. Collaboration is a form of copying: you copy from a fellow student (who you might be working with during a practical class, for example) rather than copying from a published source. This is often seen by students as a legitimate practice, but it is important to remember that whilst you will often be encouraged to work with a fellow student to conduct a practical experiment or undertake fieldwork, this doesn't usually mean that you are being asked to collaborate on the writing-up aspect of the report as well. Obviously there will be similarities between two reports written by two students who had worked together to conduct an experiment, but these will largely be confined to the methods and results sections. However, in the discussion section, the students will be expected to analyse and draw conclusions from their findings *independently* of each other. For example, compare the two discussion sections in Figure 8.2 from students A and B who were lab partners for a practical class on electrical activity in skeletal muscles.

FIGURE 8.2 Collaboration in lab write-ups.

Student A	Student B
Discussion In these experiments, observations were made of the changing patterns of electrical activity in skeletal muscles during voluntary contractions. During contractions of progressively increasing force, there was an increase in the electrical activity recorded from the muscle. This increase was related to the two different ways in which the nervous system controls muscle force: an increase in the action potential frequency of individual motor units and the recruitment of additional motor units (Berne and Levy, 2000). Observations of single unit activity showed that the firing began at about 8 impulses/s, increasing to 40 impulses/s at maximum force . . .	**Discussion** Recordings have been made of the patterns of neural activity in skeletal muscles during the course of voluntary contractions. When the force was progressively increased, there was an increase in the electrical activity recorded from the muscle. This increase was attributed to the two ways the nervous system regulates muscle force, namely by increasing the action potential frequency of individual motor units and by recruiting additional motor units (Berne and Levy, 2000). Recordings of the activity in single motor units showed that the discharge began at about 8 impulses/s and increased to 40 impulses/s at the maximum force . . .

You can see that there are definite similarities between the two discussions; the marker would probably conclude that the students had collaborated on the write-up of the discussion section and had therefore plagiarized.

Now look at the examples of students A and C in Figure 8.3.

FIGURE 8.3 Independent lab write-ups.

Student A	Student C
Discussion In these experiments, observations were made of the changing patterns of electrical activity in skeletal muscles during voluntary contractions. During contractions of progressively increasing force, there was an increase in the electrical activity recorded from the muscle. This increase was related to the two different ways in which the nervous system controls muscle force: an increase in the action potential frequency of individual motor units and the recruitment of additional motor units (Berne and Levy, 2000). Observations of single unit activity showed that the firing began at about 8 impulses/s, increasing to 40 impulses/s at maximum force . . .	**Discussion** The aim of the experiments was to use electromyographic (EMG) recordings from the whole hypothenar and from single motor units to investigate the neural control of muscle force during voluntary contractions. The initial test contraction involved a controlled increase in force and the recordings showed a progressive increase in the level of EMG activity paralleling that of the force production. It was predicted that this increase in EMG activity was made up of two different elements of neural control: variation in the rate of action potential discharge in the individual motor units and recruitment of additional motor units (Berne and Levy, 2000), these two processes occurring in parallel. To test this prediction, recordings were made of the activity of single motor units . . .

Although we would still expect to see similarities in the methods and results sections, we can see that the students have thought independently about their findings and have come to their own conclusions. This doesn't mean (in this case) that they have come to different conclusions, but the wording and sentence structure suggests that they have thought for themselves.

Self-plagiarism

It is also possible to self-plagiarize. This is when students submit their own work, which they have previously submitted for another part of their course, but fail to declare that it had been used for such a purpose. This is probably the type of deliberate plagiarism that students find the most frustrating; 'I'm the author of the work' you might reason, 'so surely I can submit it as my own work for whatever aspect of my course I choose'. It may be difficult not to have some sympathy with this view, but nevertheless self-plagiarism is considered unethical because of the deception involved in submitting the same material for credit in different modules or courses. On a more positive note, writing can always be improved upon, so it is in your own interest to do more than just recycle previously submitted work; it also helps the learning process. This is in addition to the more obvious point that a good answer written for one essay question, for example, is probably not going to be specific enough to be considered a good answer to a different, even if related, essay question. In short, although disallowing self-plagiarism may seem a little harsh, it is to your advantage to tackle each piece of assessed work as a separate exercise; allowing you to tailor your answer to the specific task and broaden your knowledge of the subject.

Buying

It is possible, on an apparently increasing number of websites, to buy essays, even complete dissertations, for submission as assessed work. This is the most blatant and outrageous form of plagiarism there is. Just in case you are ever tempted to do this, here are a number of good reasons why you shouldn't:

- it's expensive;
- it's fraudulent;
- you learn nothing about the subject;
- it will be easy for your tutors to spot if there is a significant disparity between a bought essay and your usual standard of work.

Don't do it!

8.1.4 Accidental plagiarism

Some plagiarism is deliberate, but more commonly it is accidental, or at least less deliberate than the examples cited above. There are lots of reasons for accidental plagiarism; we deal with some of them below, and use them to form the basis for the following section: *How to avoid plagiarism.*

Ignorance

Not that you will have this excuse after reading this chapter, but ignorance is the most basic reason. 'I didn't know' is a common response to accusations of plagiarism, particularly from undergraduates in the early stages of their degree course.

Poor planning

Poor planning or failing to manage your time well may seem unrelated to plagiarism, but they are, in fact, frequent causes of it. Most students would be shocked at the idea of deliberate plagiarism, especially in a premeditated manner, but when you are running out of time to complete an essay the temptation to take short-cuts becomes stronger. Read the case study below and see if you can appreciate what we mean.

Case study: Steve

> Steve was busy with another assignment and then went out to celebrate when it was finished. Waking up in the morning he remembered that he had a second assignment to hand in and so quickly used Google to search for some information on the web. Using a mixture of copying, pasting, and rephrasing he put together his assignment in a couple of hours. However, his tutors spotted what he had done and he was given a mark of zero.

So plagiarism as a result of poor planning is more plausible than it may at first seem.

Failure to record source details

Failure to record source details is probably the simplest form of plagiarism. It starts at the note-making stage; you find an appropriate piece of text, image, or data that you would like to reproduce or adapt in some way for your essay or report, but you don't write down where the material came from. At the essay-drafting stage you either forget you got it from somewhere else and believe it to be your own original work, or you realize it came from elsewhere but either can't be bothered or don't have the time to re-trace the source, so you use it anyway without referencing and hope that no one will notice. Does this seem implausible? Read the following case study and see if you understand what we mean.

Case study: Andria

> Andria took a lot of notes while she was researching her assignment, but didn't pay much attention to noting down the referencing details of her sources. When she wrote her assignment she ended up very unsure about what were her own ideas and what she had taken from books. Her lecturer is now talking about plagiarism but she feels that she was only a bit disorganized and is horrified that she is being accused of dishonesty.

Plagiarism due to failure to record source details is, therefore, a very real problem.

Inappropriate notes

Inappropriate notes are also a common cause of plagiarism. It is reasonably straightforward to record source details (in theory at least) but how do you know what sort of notes to make on the source? See if the following case study helps you understand what we mean.

Case study: Abdul

> Abdul has a good understanding of the course; however, he finds it difficult to put things into his own words. Academic text books always seem to put things so well that he hasn't got much to add. Sometimes it just seems easier to copy out chunks of different books, paraphrase them a bit, and organize them into an answer. He doesn't feel that this should be considered plagiarism as he has researched the material and put the assignment together from a range of different sources. Unfortunately his tutors don't agree.

Academic text books or journals can indeed 'put things so well' that it's difficult for a student to know what to add. This is why it is so important to work at understanding the concept that you are dealing with, and it takes a certain amount of confidence to do this. We will deal with this in the *How to avoid plagiarism* section.

Incomplete citing and/or referencing

Sometimes students inadvertently commit plagiarism by simply not recording their citations and references fully. This could be due to a lack of understanding of what is required, or a lack of willingness to add what are sometimes just considered to be 'finishing touches', but are in fact crucial to the integrity of their work. Again, see if the following case study helps to illustrate this.

Case study: Jasmine

> Jasmine worked hard on her essay and was pleased with what she produced. However, she has never understood the point of the referencing system. She included all of the books that she used in a reference list at the end of the essay but didn't bother with citing her sources in the body of the essay. She thought the reference list at the end would be enough but, unfortunately, her lecturers viewed this lack of citing as plagiarism. They covered the assignment in red ink and gave her a very low mark.

So, although sorting out the citations and references might just seem like an optional extra, it is vital to get it right. Referencing correctly is dealt with in section 8.2.5, and more detailed guidance is available in the section on *Citations and references* in Chapter 3, *Working with different information sources*.

Lack of engagement with sources and failure to express own ideas

This may possibly be the reason that requires the most thinking to address: it is where a student may well have planned ahead, recorded source details, made appropriate notes, and cited and referenced accurately, but still doesn't demonstrate that they have thought a great deal about what they have written. It is perfectly possible, in an assessed piece of work, simply to string together lots of other authors' ideas (either quoted or paraphrased); but if you fail to include your own ideas, in the form of comment or analysis on the material, you still might be plagiarizing, even if the source material is cited and referenced correctly.

8.2 How to avoid plagiarism

We have dealt with what plagiarism is, having thought about its definition and how academic honesty (or not plagiarizing) underpins the academic method. We have also considered the main reasons students plagiarize, identifying both deliberate and accidental forms of plagiarism. We are now going to consider how to avoid it.

8.2.1 **Know what plagiarism is**

The first step to avoid plagiarism is to be able to recognize it when you see it. This is different from simply defining it; can you tell in practice when something is or is not plagiarized? Use the following exercise to test your knowledge so far.*

Essay extract exercise

Original text:

> During the last 60 years the development of effective and safe drugs to deal with bacterial infections has revolutionized medical treatment, and the morbidity and mortality from microbial disease have been dramatically reduced (Rang *et al.*, 1999).

Essay extract 1:

> During the last 60 years the development of effective and safe drugs to deal with bacterial infections has revolutionized medical treatment, and the morbidity and mortality from microbial disease have been dramatically reduced.

Is this extract plagiarized, yes or no?

Essay extract 2:

> During the last 60 years the development of effective and safe drugs to deal with bacterial infections has revolutionized medical treatment, and the morbidity and mortality from microbial disease have been dramatically reduced (Rang *et al.*, 1999).

Is this extract plagiarized, yes or no?

Essay extract 3:

> 'During the last 60 years the development of effective and safe drugs to deal with bacterial infections has revolutionized medical treatment, and the morbidity and mortality from microbial disease have been dramatically reduced' (Rang *et al.*, 1999).

Is this extract plagiarized, yes or no?

* Exercise adapted from Willmott, C.J.R. and Harrison, T.M. 2003. An exercise to teach bioscience students about plagiarism. *Journal of Biological Education* **37**, 139–140. (Thanks to the *Journal of Biological Education* for permission to adapt this article.)

Essay extract 4:

> In the 4th edition of their textbook Pharmacology (1999), Rang, Dale and Ritter state that: 'During the last 60 years the development of effective and safe drugs to deal with bacterial infections has revolutionized medical treatment, and the morbidity and mortality from microbial disease have been dramatically reduced.' Such a bold assertion understates the ongoing threat posed by microbial infection. It is estimated, for example, that worldwide there were over 8 million cases of tuberculosis in 1998 (WHO, 2000).

Is this extract plagiarized, yes or no?

Essay extract 5:

> The development of safe and effective drugs to deal with bacterial infections has dramatically reduced the death rate arising from microbial diseases (Rang *et al.*, 1999).

Is this extract plagiarized, yes or no?

Essay extract 6:

> During the post-war years, the development of effective and safe drugs to deal with bacterial infections has transformed medical treatment, and death and illness resulting from microbial disease has been dramatically reduced (Rang *et al.*, 1999).

Is this extract plagiarized, yes or no?

Essay extract 7:

> The availability of antimicrobial compounds has transformed healthcare in the period since the Second World War. People are far less likely to die or even be seriously ill than they had been prior to the introduction of these drugs (Rang *et al.*, 1999).

Is this extract plagiarized, yes or no?

Essay extract feedback

Essay extract 1:

This is copied word-for-word but there is no reference. It is a definite case of plagiarism.

Essay extract 2:

This is marginally better than extract 1 because the source has been acknowledged by the inclusion of the reference. However, it is still plagiarism: the text is copied word-for-word and should be in inverted commas to indicate that it is the exact wording of the source.

Essay extract 3:

Inverted commas are used to acknowledge that words have come from a different source. The inverted commas here make it clear that the student is acknowledging that both the ideas and the words have come from the textbook, so it is not plagiarized. However, stringing together a series of quotations does not demonstrate your understanding of the subject and so is likely to score low marks (we will deal with this later in section 8.2.6 on *Use sources appropriately and express own views*).

Essay extract 4:

The quotation is indicated and used in an appropriate way; it has been commented on by the author and supplemented with further information, and so is not plagiarized.

Essay extract 5:

This is only a cosmetic alteration. The wording and sentence construction bears very close resemblance to the source and so is plagiarized.

Essay extract 6:

This student has just swapped a few words, perhaps using a thesaurus to find replacements, but this is not sufficient to be considered new work. The thinking has gone into finding alternative words to avoid directly quoting rather than into understanding the statement, and so is plagiarized.

Essay extract 7:

The student has understood the source and put the relevant information into his or her own words. This demonstrates the student's engagement with the text and ability to explain the information relevant to the essay, and so is not plagiarized.

8.2.2 Plan ahead

We have already noted that bad planning, and the pressures it creates, increases the temptation to plagiarize. So whilst bad planning is not a direct cause of plagiarism, it is still a significant factor nonetheless, and so needs addressing.

The solution to bad planning is, of course, good planning. This strikingly obvious statement is worth mentioning, as it highlights the fact that the skills required to address this issue are simply common sense. However, there can be a world of difference between knowing what you should do and actually doing it. In Chapter 5, *Writing essays*, the different stages involved in writing an essay were dealt with. It is helpful to think of these stages not just in terms of separate tasks but also as distinct periods of time that need planning. Chapter 5 recommended that you think about the following stages when writing an essay:

- analyse the question and brainstorm ideas;
- research the topic;
- write a plan;
- write first draft;
- review and re-draft;
- proof-read (including checking of citations and references).

A good way to plan ahead is to order these tasks into some sort of timeline, calendar, or diary. Figure 8.4 shows an example timeline representing a period of 3 weeks that you might have available to write an essay. It is worth spending a short amount of time allocating your tasks in this way to provide some structure and discipline to your time and so hopefully reduce the chances of you leaving things to the last minute, which would, as we have seen, increase the chances of you plagiarizing.

This method also has the advantage of making the task less daunting. If you had on your 'to do' list 'write ion channel essay', it would probably be a long time before you were able to cross it off, which can be quite de-motivating. However, if your 'to do' list said things like 'gather journal information for essay' and 'write essay plan', these are things that would take hours rather than weeks so you would be able to cross things off much sooner. This makes it easier to discern progress and so is motivating rather than de-motivating; you feel like you are getting somewhere.

Note also in Figure 8.4 that there is a significant gap between 'write first draft' and 'review and re-write first draft'. This is important for a couple of reasons: firstly, you could reward yourself in some way for getting this far in the process, again a motivating factor; and secondly, coming back to an essay after a bit of a break enables you to be more objective about what you have written. This objectivity increases the likelihood that you will be able to make significant improvements.

FIGURE 8.4 Essay tasks timeline.

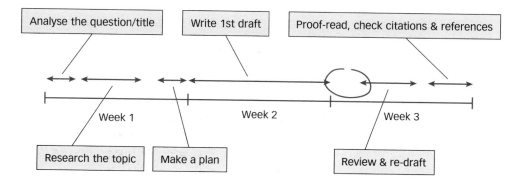

8.2.3 **Record source details**

As we have seen, it is important to record source details if you don't want to plagiarize. It sounds simple enough, but, as with planning ahead, when the pressure is on it is both easier and more tempting to make mistakes.

Imagine you had made this note during a study session in the library last week:

> Ion channels are not simply tubes that span membranes through which ions can rapidly flow. Instead, they are highly sophisticated molecular machines that respond to chemical and physical changes in their environments and undergo precisely timed conformational changes.

It is a good piece of writing and perfect for your essay on ion channel function. The only trouble is you can't quite remember if these are your words or someone else's because you didn't write down any more information; an easy mistake to make.

The words are actually taken from *Biochemistry* by Berg *et al.* (2002). So if you were to use them in your essay without putting them in inverted commas ('. . .'), citing, and referencing them, you would be plagiarizing. The good news is that, because failure to record source details is a simple mistake to make, there is also a simple solution. It just takes a bit of discipline.

There are four main sources that you will use in your studies: books, journals, websites, and lectures. Table 8.1 summarizes the information you need to record for each one in order to reference fully.

All you need to do is make sure you record the source details relevant to the type of source before you begin to make notes. It's as simple as that. See Table 8.2 for examples of citations and references from the source details.

8.2.4 **Make appropriate notes**

Recording source details, then, is quite straightforward, as long as you know what details you need to record for which source. But what about when you actually need to make notes on the source material? How do you do it?

When making notes it is important to ask yourself a simple question: what am I making notes for? There are several possible reasons you might have. You might be making notes to:

- select information;
- understand information;
- remember information.

Clearly these three areas overlap considerably, but in order to keep things simple we are going to deal with them separately. Chapter 3, *Working with different information sources*, deals with making notes to understand information and the reading techniques necessary to achieve this efficiently; Chapter 13, *Getting the most out of revision*, deals

TABLE 8.1 Source details to record for different information types

Source type	Source details required
Books	Name(s) of author(s)*
	Year of publication
	Title of book
	Name of publisher
	Place of publication
	Page reference
	*If it's a chapter of an edited book you also need to include the 'Editor(s) name(s)', 'Title of chapter' and 'First and last page numbers of chapter'.
Journals	Name(s) of author(s)
	Year of publication
	Title of paper
	Title of journal
	Volume number
	First and last page number
Websites	Name(s) of author(s)
	http://address
	Title
	Date page accessed*
	*Often internet pages indicate when they were last updated—this is also important information to record if available.
Lectures	Name of lecturer
	Name of lecture/lecture series
	Date of lecture
	Location (e.g. University of Leicester)

with taking notes to remember information. Here we will just focus on taking notes to *select* information.

At the beginning of this chapter, when we were thinking about the different elements involved in the definition of plagiarism, we identified that 'someone else's work' could be their words but also their ideas, data, or images. When making notes, therefore, we are not merely thinking about text, we also need to remember other forms of information too: including ideas, data, and images. In simple terms, you have two options when considering how to include another person's work in your own: you could reproduce it exactly (so in the case of text this would mean quoting it word-for-word) or you could adapt it in some way (again, in the case of text this would mean putting it into your own words, i.e. paraphrasing it).

TABLE 8.2 Examples of source details recorded

Source type	Source details required	Example source details	Example citation	Example reference
Books	Name(s) of author(s): Year of publication: Title of book: Name of publisher: Place of publication: Page reference:	Dawn Hawkins 2005 Biomeasurement OUP New York pp. 95–96	Hawkins (2005)	Hawkins, D. 2005. *Biomeasurement*, OUP, New York, pp. 95–96
Journals	Name(s) of author(s): Year of publication: Title of paper: Title of journal: Volume number: First and last page number:	Watson, J.D. and Crick, F.H.C. 1953 Molecular structure of nucleic acids; a structure for deoxyribose nucleic acid Nature 171 737–738.	Watson and Crick (1953)	Watson, J.D. and Crick, F.H.C. 1953. Molecular structure of nucleic acids; a structure for deoxyribose nucleic acid. *Nature* **171**, 737–738
Websites	Name(s) of author(s): http:// address: Title: Date page accessed:	Heitler, W.J. http://www.standrews.ac.uk/~wjh/jumping/ How grasshoppers jump 27/06/08	Heitler (2008)	Heitler, W.J. 2008. How grasshoppers jump. [online] available from: http://www.standrews.ac.uk/~wjh/jumping/ [accessed 27 June 2008]
Lectures	Name of lecturer: Name of lecture/series: Date of lecture: Location:	Jon Scott Physiology of locust hearing 23/11/07 University of Leicester	Scott (2006)	Scott, J. 2007. Physiology of locust hearing [lecture 23 November 2007].

For example, if you were writing an essay on 'Factors influencing the development of cancer' and found some useful information in *Introduction to the Cellular and Molecular Biology of Cancer* by Knowles and Selby (2005), you may decide that the exact wording and/or imagery is important to you and so want to reproduce it exactly, as follows:

Text:

> 'There are many factors which influence the development of cancer. These include both endogenous factors such as inherited predisposition and exogenous factors such as exposure to environmental carcinogens and infectious agents. Another factor which has a clear influence on the type of cancer which develops is age' (Knowles and Selby, 2005).

Image:

FIGURE 8.5 Colon cancer incidence in men in England in 1999 (reproduced from Knowles and Selby (2005), Figure 1.3, p. 8).

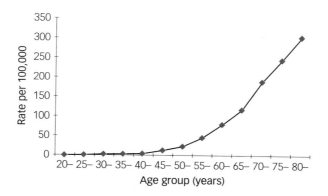

However, if it was the *idea* rather than the exact wording or imagery that was important, you would need to adapt it in some way (which may mean simplifying or expanding on it), as follows:

Text:

There are many reasons for the development of cancer, but most can be grouped into two broad categories; those originating from within the organism (e.g. genetic predisposition) and those having an external origin (e.g. carcinogens). A third, but no less important, category is age (Knowles and Selby, 2005).

Image:

FIGURE 8.6 Colon cancer incidence in men in England in 1999 (adapted from Knowles and Selby, 2005).

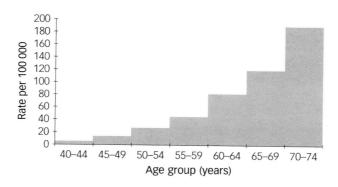

The important thing to remember is that whether you are reproducing another's work (e.g. quoting text) or adapting it (e.g. paraphrasing text), both need referencing. In the case of quoting, the quoted text is placed in inverted commas ('. . .') (to make clear that you are using someone else's words verbatim) followed by the citation; in the case of paraphrasing the paraphrased text is followed by the citation.

8.2.5 Reference correctly

As with planning ahead and recording source details, referencing correctly is simple enough in theory, you just need to pay attention to the details. Here are some simple steps that should help (more detailed guidance is available in the section on *Citations and references* in Chapter 3, *Working with different information sources*).

Check what style of referencing is expected from you

Different institutions have different styles of referencing; the most common system in the biosciences is an author–date system (see Chapter 3) but some will use footnotes, some will use endnotes, some will use variants of these systems, and there are many others besides! Even within an individual institution (e.g. a university) it is unlikely that all departments use the same type of referencing style. In fact, there are likely to be minor variations even within departments. This can create a lot of confusion and it can be very tempting to decide not to bother with the details. However, when it comes to referencing, the details are important.

Know the difference between citing and referencing

An important distinction to make is the difference between citing and referencing. In general terms **referencing** refers to the practice of acknowledging your sources. More precisely, however, referencing in an academic context has two elements:

- the citation in the text;
- the full reference at the end of the piece of work (e.g. your essay or practical report).

The **citation** is simply the way you refer to another's work in your own work; in the case of the author–date system this means the text, idea, image, or data followed by the surname of the author (or authors) and the date of the publication. For example, 'The novel feature of the structure is the manner in which the two chains are held together' (Watson and Crick, 1953). The **reference** is the full details of the authors and publication at the end, e.g. Watson, J.D. and Crick, F.H.C. 1953. Molecular structure of nucleic acids; a structure for deoxyribose nucleic acid. *Nature* **171**, 737–738. See Figure 8.7.

FIGURE 8.7 The relationship between citations and references.

'The novel feature of the structure is the manner in which the two chains are held together'
(Watson and Crick, 1953).

References
Watson, J.D. and Crick F.H.C. 1953. Molecular structure of nucleic acids; a structure for deoxyribose nucleic acid. *Nature* **171**, 737–738.

Use *et al.* correctly

Et alii is a Latin term meaning 'and others'. It should therefore be written in italics (because it's Latin) and be followed by a full stop (because it's an abbreviation). It is used when you are citing a source which has more than two authors, for example, when referring to the book *Molecular Biology of the Cell* published by Garland Science, instead of having to write Alberts, B.; Johnson, A.; Lewis, J.; Raff, M.; Roberts, K.; Walter, P. (2002), you would simply write, Alberts *et al.* (2002), which, as you can see, is very handy.

List the references in alphabetical order at the end

Your reference list at the end of your document should be in alphabetical order by surname of the author, for example:

Dalgleish, G. 2002 . . .

Harrison, W. 1999 . . .

Wells, B. 2004 . . .

If there is more than one author for a publication then it is the surname of the first author that counts, for example:

Cann, T. 2005 . . .

Jones, M. and Adams, C. 2003 . . .

If there are two authors with the same name or one author with two publications, then you need also to order the list by date, for example:

> Dalgleish, R. 2002 . . .
> Dalgleish, R. 2004 . . .

If an author has more than one publication in a given year then the list needs to distinguish between these, usually by including letters in the dates, for example:

> Robbins, H. 2004a . . .
> Robbins, H. 2004b . . .

This distinction would also need to be reflected in the citation, so (Robbins, 2004a), and so on.

Consider using bibliographic software, if available

If you are going to be using a large number of references, for instance in a final year dissertation, you may wish to use bibliographic software. Bibliographic software is designed to allow you to input, organize, manage, retrieve, and format lists of references more easily than manual input. Two examples are EndNote and RefWorks, but it will depend on what software your university subscribes to. Whatever system your university subscribes to it is worth finding out more about it as it may save you a lot of time. However, even if you use such software, you still need to understand the structure of references if you are going to use it appropriately.

8.2.6 Use sources appropriately and express own views

We noted in section 8.2.4 above that 'In simple terms you have two options when considering how to include another person's work in your own'—reproducing exactly or adapting for your own purposes. These are only simple terms because merely using another person's work and referencing it correctly is only a starting point. If an essay or report were to be just a collection of other people's ideas it would not score very good marks; in order to do well you need to engage actively with source material and express your own views by providing comment or analysis on your sources. This is what this final section in *How to avoid plagiarism* is about.

There are many ways in which you can engage with source material. We are going to look at three examples. Engaging with source material to:

- illustrate a point;
- provide evidence;
- contrast arguments.

Precisely which method of engagement you choose is somewhat secondary; the important point is that you must engage with the sources you use in some way rather than using material without comment or analysis.*

Illustrating a point

One way to use source material is to use it to illustrate a point. See Figure 8.8 for an example from the field of genetics. In this example, the idea is introduced, the evidence is summarized, and then an explanation is given of what this evidence illustrates.

FIGURE 8.8 Using sources to illustrate a point.

> More than four decades ago Marshall Urist began a series of experiments designed to characterize the factors necessary to induce bone formation (Urist, 1965). In 1975 he demonstrated that the 'bone morphogenetic factor' was part of the protein component of bone, rather than being associated with the hard mineralized fraction (Urist *et al.*, 1979). However, he was initially unable to isolate the factor as a single protein. This illustrated that the induction of bone growth might be controlled by several co-acting protein factors.

Introduction

Summary of evidence

Explanation of what this illustrates

Providing evidence

Another way to use source material is to use it to provide evidence. See Figure 8.9 for an example, again from the field of genetics. In this example, the idea is introduced, the evidence is summarized, and then the evidence is explained.

FIGURE 8.9 Using sources to provide evidence.

> Over several years, Urist continued to characterize his bone morphogenetic factor, making improvements to the purification procedures (Urist *et al.*, 1984). He showed that de-mineralized bone could be fractionated to yield low molecular weight protein morphogenetic components, but none of these was capable alone of inducing bone formation (Urist *et al.*, 1987). These studies suggest that bone morphogenesis is a complex process requiring the interaction of several factors.

Introduction

Summary of evidence

Explanation of evidence

*Thanks to Dr Raymond Dalgleish from the Department of Genetics, University of Leicester, for these examples.

Contrasting arguments

A third way to use source material is to use it to contrast arguments. See Figure 8.10 for an example, again from the field of genetics. In this example, the idea is introduced, some evidence is paraphrased, this evidence is criticized, contrasting evidence is then introduced, and finally the writer's opinion on the issue is offered.

FIGURE 8.10 Using sources to contrast arguments.

For many years, the precise identity of Urist's bone morphogenetic protein (BMP) remained unresolved except for the fact that it had a major component with a molecular weight of 18 kDa. In 2005 Behnam *et al.* demonstrated that the amino acid sequence of peptides derived from highly purified BMP show identity to regions of a previously characterized protein known as spp24. However, spp24 has a molecular weight of 24 kDa. It could be argued that Behnam *et al.* reached a mistaken conclusion because of failures in the strategy used to characterize BMP or that they simply failed to fully explain their findings in the published account of their work. Subsequent experiments, involving recombinant protein expression, have shown that the spp24 protein undergoes post-translational cleavage, yielding a product consistent with Urist's original 18 kDa BMP.

Introduction

Paraphrasing

Criticism

Speculation about the discrepancy

Resolution of the discrepancy

The above three ways of engaging with source material are not exhaustive; there are many others including agreeing, disagreeing, synthesizing, reconciling, and developing. The important thing is that you provide some comment or analysis on the sources you use, rather than simply record them.

8.3 **Seven good reasons not to plagiarize**

As we have seen, avoiding plagiarism is not just about correct references, it is an unavoidable part of good academic writing. In case you need any more convincing, here are seven good reasons for you not to plagiarize.

1. **Plagiarism short-circuits learning**

   When you are set an assessed piece of work it is easy to focus on the tangible end-point, that is, the mark you are awarded. However, your tutors set you work, not merely as a means of allocating marks, but also as a means of helping you to learn.

The research you conduct, the selecting of information, and the process of writing all aid the learning process. The more you plagiarize, the less you learn.

2. Plagiarism destroys independent thinking

Your tutors are interested in what you think about a subject. Obviously, your point of view needs to be supported by appropriate evidence, but, as we have seen, that evidence needs to be accompanied by your own comment or analysis; your own ideas and conclusions. If you plagiarize by not commenting or analysing evidence, you are not thinking for yourself; you are merely communicating the ideas of others and have failed to think independently.

3. Plagiarism is unfair to your peers

How would you feel if a student on your course scored a higher mark than you on an essay but you knew that they had cheated? Make sure you don't make anyone else feel the same way about you.

4. Plagiarism is a form of theft

Words, ideas, data, and images that have been written, thought of, discovered, or created by someone else belong to that person. Using them without acknowledging the source is stealing.

5. Plagiarism can result in severe penalties

Different universities have different penalties for students who have plagiarized. The type of penalty depends on the precise nature of the plagiarism, the extent to which the work contributes to final marks, and whether or not it is a first offence. Common penalties include:

- an awarding of zero for the piece of work, the requirement to repeat the work, plus an official warning;
- an awarding of zero for the module and withdrawal of the right to re-sit;
- the downgrading of degree class by one division;
- expulsion from the course.

6. Plagiarism is now more difficult to get away with

Many universities are now using plagiarism detection software. Students are required to submit their work electronically, the software then compares the work with a vast array of other sources, including core texts, journal articles, internet sites, and previously submitted student work. A report is produced highlighting the areas of similarity with the compared sources (Figure 8.11). The tutor would then interpret the report to decide whether or not it is plagiarized. The software is becoming increasingly sophisticated and makes getting away with plagiarism much more difficult. However, even if your university doesn't currently operate an electronic detection system, it is worth noting that academics are also very capable of quickly spotting plagiarism using more old-fashioned methods.

FIGURE 8.11 Example of plagiarism software originality report generated by Turnitin.

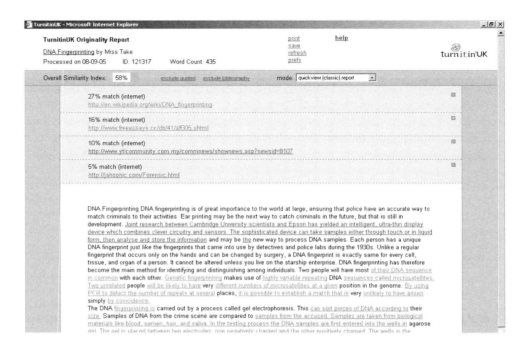

7. **Not plagiarizing means your learning will be more effective**

Finally, remember that avoiding plagiarizing is not simply about evading punishment. The skills that help you not to plagiarize are the skills that enable you to be an effective learner. Learning these skills will help you in all areas of your academic work; helping you to understand your subject better and so help improve your grades.

✳ Chapter summary

As we have seen, plagiarism is an important issue, and you need to know both what it is and how to avoid it if you are going to do well in your studies. Planning, recording source details, making appropriate notes, referencing correctly, using sources appropriately, and expressing your own views are all important elements. If you can address these issues you won't simply be avoiding plagiarism, you will become a better scholar.

Chapter 9

Preparing scientific presentations

⮕ Introduction

The ability to present information verbally to a group of people is an increasingly important skill for students undertaking bioscience degrees, or any other kind of degree for that matter. Some students find the prospect of standing up in front of a group of their fellow students and academic staff to give a presentation is the stuff of nightmares, even more so when the presentation is assessed. For other students presentations are not quite such a frightening prospect, but, even so, the public nature of a presentation, as compared with the privacy and anonymity of an essay or exam, for example, is a factor that makes getting presentations right all the more important; no one likes being embarrassed in front of their peers.

In this chapter we will begin by considering the purpose of presentations as a form of assessment; we will then identify some of the characteristics of effective presentations, and scientific presentations in particular. We will look at the importance of finding out the answers to a number of key questions before progressing on to analysing the brief, researching the topic, and planning and ordering your presentation's content. Finally, we will highlight the importance of preparing effective and appropriate notes. This chapter complements Chapter 10 which looks specifically at delivering, as opposed to preparing for, your scientific presentation.

9.1 What are presentations supposed to achieve?

It can be useful to think about why you are being asked to undertake a particular form of assessment, as different forms of assessment are chosen for different reasons. The assessment of presentations is usually divided into two broad areas: content, and the communication of the content (or 'substance' and 'style'). Both of these elements are important and you need to be careful not to focus on one at the expense of the other.

For example, it is possible for your presentation to have relevant and informative content, but if you don't communicate it in a well-structured and engaging manner you would score relatively low marks in an assessment. Equally, you should avoid trying to dress up content that you don't understand and haven't put much effort into researching by giving a slick presentation with a few impressive PowerPoint slides. If you do this it is usually fairly obvious to the audience (and it certainly will be to your tutor) that your presentation is all style and no substance. So, both substance and style are important.

We will consider the following brief for a sample presentation:

> Prepare and deliver a five-minute oral presentation on the use of global positioning systems (GPS) in the study of migrating populations. The presentations will take place at 12 noon on Friday 1st March in room 101.

We can identify a list of the skills being exercised in order to complete the task as follows:

- using feedback from previous presentations to produce a better piece of work;
- analysing the question or title (and associated instructions);
- researching the topic;
- bringing together the information gathered from a range of sources, e.g. lecture notes, textbooks, research papers, web articles, etc.;
- planning the information to address the brief and make sure that all the key points are covered;
- matching the level of detail against the type of audience you will be addressing;
- ordering the material in a logical manner;
- creating visual aids to support your key points;
- preparing appropriate notes so that you can remember what you want to say;
- practising your presentation beforehand so that you become familiar with communicating the content;
- delivering your presentation in a relaxed and engaging manner;
- referencing the information correctly to acknowledge the original sources;
- planning and managing your time to make sure everything is ready by the deadline.

As was similarly identified in Chapter 5, *Writing essays*, this is a surprisingly long list and could probably be extended. The important point is that while some of these skills, such as researching the topic and preparing your notes, are obvious and easily recognized as part of the process, others, such as using feedback and referencing information, are more implicit but still very important if you are to plan and deliver a good presentation. As we go through the process of preparing a presentation (this chapter) and delivering a presentation (Chapter 10), we will consider these different skills and give you guidelines for improving the quality of your scientific presentations.

9.2 **What makes a good presentation?**

What, then, makes a good presentation? Nearly everyone has an opinion on this question. Even if you have never given a presentation before, you will have seen others give presentations (for instance a lecture), and you will probably remember some that were really good, and some that were really bad! It is useful to think about what makes a good presentation as it will give you a list of things you want to try to copy and things you want to try to avoid.

9.2.1 **Characteristics of effective presentations**

Obviously, not all presentations should be the same. In fact, it is important that as you gain experience in presenting you develop your own style, rather than trying to conform too rigidly to a set standard. However, effective presentations usually exhibit certain characteristics. These include that the presentation is:

- relevant;
- informative;
- well structured;
- engaging.

This list is not exhaustive, though these are important characteristics of effective presentations. You should also recognize that these characteristics are not self-contained; they overlap and are dependent on each other. To go back to our substance and style distinction that we mentioned above, relevance and informativeness are largely to do with substance, whereas structure and engagement are largely to do with style. We will deal with each of these characteristics in turn.

Relevant

The starting point for any presentation is the presentation brief, as it is this that will govern the content and focus of the presentation. Although it can be tempting to jump straight from a brief to the bits you find interesting about the topic or the bits that you happen to have found some useful information about, it is important that you keep going back to the title and instructions you have been given to make sure that your material is relevant, just like you should do for an essay. But a presentation should not only be relevant with regard to the brief; it should also be relevant with regard to the audience. Are you pitching your presentation at the correct level for the audience in terms of their current understanding and their ability to assimilate the material? For instance, is it likely that you will be using quite technical terms on occasions? If so, is your audience already familiar with these terms, or do you need to define them? So, a presentation needs to be relevant, both to the brief and to the audience.

Informative

Not only should a presentation be relevant, it should also be informative. What is the content that you want to communicate to your audience? What do you actually want to say? It can be very helpful during your preparation to try to summarize the content of your presentation in a single sentence or three or four short bullet points. For instance, to go back to our example above, you could summarize the content of that presentation as follows: 'Short explanation of what GPS is, how the technology has been applied to study migrating populations, and examples of Wildebeest and Canada geese.' If you can't summarize the content in a sentence or two then it probably isn't focused enough. So, a presentation needs to be informative.

Well structured

Now we start to get on to the style, as opposed to substance, elements. The structure of a presentation is a good example of where there is significant overlap between style and substance: good structure begins with the content. However, you need to be able to communicate that structure to the audience so they can follow what you are saying. If a presentation is easy to follow, it will in part be due to its good structure. There are two simple ways of communicating the structure of your presentation to your audience: you could tell them or you could show them (or both).

Telling the audience the structure involves outlining it verbally at the beginning, for instance:

> This presentation is divided into three main sections. Firstly, I'll give a brief explanation of what GPS is. Then I'll explain how the technology has been applied to the study of migrating populations. Finally, I'll tell you about two specific examples: Wildebeest and Canada geese.

In addition to outlining the structure of the presentation at the beginning, it will also be helpful to give verbal cues to the audience as you go along so they can clearly follow the progression of the presentation. Verbal cues act like signposts to the audience so they know where you are going. Some of these verbal cues include:

- a single word such as 'Firstly', 'Secondly', and so on;
- linking statements such as 'So that's the background to the technology, let's now think about its application . . .';
- concluding statements such as 'In conclusion . . .' and 'Finally . . .'.

That might all seem fairly obvious to you, but it is surprising both how often these signposts are neglected and how helpful it is for the listener when presenters make a point of using them. As you become more experienced in presenting you will give these verbal cues much more naturally, but for as long as they don't come naturally it is a good idea to make a conscious point of using them.

Showing the audience the structure could simply be a case of producing an outline slide that summarizes the main points. An example is shown in Figure 9.1. You could use this at the beginning when you tell the audience about the structure. If the structure was particularly important or particularly complex, you might also repeat the outline slide within the presentation when you mention the verbal cues. However, it is possible to give a presentation so much structure that it becomes very separated out and doesn't flow, so be careful not to overdo it.

FIGURE 9.1 Outline slide showing presentation structure.

Outline
- What is GPS?
- Applications to migrating populations
- Examples
  - Wildebeest
  - Canada geese
- Conclusion

Engaging

The last characteristic of an effective presentation that we have identified is that it should be engaging; it should be interesting and hold the audience's attention. Again, there is overlap here with the content or substance aspects of the presentation (particularly the relevance to the audience), but whether or not a presentation is engaging is much more to do with how information is communicated as opposed to what is communicated. Whether or not a presentation is engaging will depend on a wide range of factors: it will need to be well paced, the presenter will need to have good eye-contact with the audience, and it will probably help if the presenter can be fairly relaxed (or at least seem to be relaxed!). All of these depend to a large extent on how much you have practised the presentation. Additionally an engaging presentation is often also helped by the appropriate use of visual aids, which we will address in Chapter 10.

9.2.2 Characteristics of effective scientific presentations

In addition to the characteristics of effective presentations highlighted above, there are certain characteristics that scientific presentations in particular should exhibit. These include that a scientific presentation should be:

- objective;
- precise;
- clear.

Again, this is not an exhaustive list and there is overlap between the different elements, but they are key characteristics nonetheless. We will address each of them, briefly, in turn.

Objective

Science is, by its very nature, objective. That is to say, observations made during scientific research should be impartial and not influenced by personal feelings or opinions. So it follows then that a scientific presentation should also be objective. This means that the content you are communicating needs to be well researched and well referenced. As was highlighted in Chapter 8, *Avoiding plagiarism*, students are usually aware that plagiarism is wrong in the context of essays, yet there is often a bit of a blind spot when it comes to presentations. But making sure that you reference the sources of information in your presentation is not simply about avoiding plagiarism; just as importantly it is about showing the objectivity of what you are saying.

Precise

Scientific presentations should also be precise. As was highlighted in Chapter 5, *Writing essays*, the use of generalizations often clouds the meaning of what is written or indicates a lack of confidence on the part of the presenter. For example, what does the following phrase actually mean?

> The animals are fairly active at night, moving around quite a bit . . .

The phrases 'fairly active' and 'moving around quite a bit' have no real meaning as their interpretation depends entirely on the listener. Relative terms such as 'quite' or 'fairly', or phrases such as 'In general terms . . .' lack precision and should, wherever possible, be replaced with more precise phrases:

> During the hours of darkness, the animals forage widely, travelling up to 2 km from the nest site . . .

Clear

The final characteristic of scientific presentations that we have identified is that they should be clear. This does not just apply to simple, uncomplicated subjects. In fact, one of the marks of a good presentation and a good presenter is the ability to explain complex things in relatively simple terms. Using simple terms should not mean that the presentation is simplistic, but it does mean that the content needs to be explained in such a way that an average member of the audience can understand it without too much difficulty.

9.3 Analyse the question or brief

We have already identified that relevance is a key characteristic of effective presentations. We observed that a presentation needs to be both relevant to the question or brief you have been set and relevant to the audience you will present to. A useful technique for analysing the question or brief is simply to write it in the middle of a piece of paper and then annotate it, as can be seen in Figure 9.2.

FIGURE 9.2 Analysing the question/title.

This technique enables you to begin to get your thoughts down on paper quickly whilst encouraging you to link your thoughts to the task you have been set, thereby helping you to keep your ideas relevant. It also allows you to make connections between ideas and to see, at a glance, how the presentation might be structured.

9.4 Research the topic

As we saw in Chapter 5, *Writing essays*, researching the topic can be a very time-consuming task. There are three things you can do to reduce, or at least control, the amount of time the research part takes.

1. Analyse the brief first so that you have some clear ideas of what you are looking for before you start looking for it. It is easy to waste a lot of time researching material that you later discover is not particularly relevant; analysing the brief provides a focus for your research.
2. Set a time limit for your research. If you don't set a time limit, the research element can just go on and on, especially if you are being particularly diligent or are very interested in the topic.

3. Look at source material in an appropriate order. If you start off with an internet search it can take you off at all sorts of tangents. If, however, you start with lecture notes and key chapters from core texts, it provides focus for your research so it is easier to discriminate when you get to the other sources (see Chapter 3, *Working with different information sources*). A suggested order is as follows:

- start with the lecture notes—check that you understand the information;
- identify the essential reading from the relevant chapters in the core texts;
- when you have mastered the basics, extend your reading by looking at books that address a particular aspect of your topic;
- look at recent research articles to bring you up to date on the subject and to add the necessary level of detail.

It is important to note that jumping to internet sources too early can actually hinder your research. You need first to establish a clear focus that is relevant to the question or brief so that you can be highly selective if you do choose to supplement your research with internet sources.

9.5 Plan and order the material

During the research stage you will have probably gathered more material than you actually need, even if you have been quite disciplined in your approach. The temptation is to include too much material because you have spent the time finding it; however, this will often result in a presentation that is difficult to follow, unfocused, or too long (or all three!). The following stages will help you to decide what material you should include.

9.5.1 Choose your main points

The starting point for selecting and organizing your material should be to choose your main points. You might not be able to list all of the main points straight away, but it is important nonetheless to make a good attempt. As you continue in your preparation you may well come back to your main points to refine or reorganize them, but you need a basic framework at an early stage to focus your thoughts and provide a structure for the rest of your material. An example structure for a 5-minute oral presentation on the use of global positioning systems in the study of migrating populations was given above, as follows:

- what is GPS?
- applications to migrating populations;
- examples: Wildebeest and Canada geese;
- conclusion.

This is a very sketchy outline, but it is enough information to get you started. A useful technique to help you draft your main points is simply to put your research notes to one side, then try to write your main points without looking at the research notes, expressing each point in a few words or a short sentence. Alternatively, you could try to articulate your main points to a friend, because expressing ideas out loud can be a good way of helping you decide whether or not they are making sense. It may also be helpful to add some notes to your outline to give you an idea of what the focus of your presentation should be. For instance, from the list above it looks like the four bullet points (or at least the first three) are equally sized sections, which probably isn't your intention. It might be better, therefore, to annotate your main points with a bit more information, as follows:

- what is GPS? (just brief background—enough for people to understand the basics);
- applications to migrating populations (how the technology has been used in this context and overview of its applicability and value);
- examples: Wildebeest and Canada geese (more detailed descriptions of the application to these specific examples);
- conclusion (summarize the main points and invite questions).

This is also the point at which you would begin to organize your main points into a logical sequence, so that each point links to, and builds on, the previous one. The sequence of your main points will be determined not only by what is logical, but also by what you are aiming to achieve. For example, if you are trying to build an argument you will want to move from background information to precise points of detail, or if you are explaining a process you might begin by explaining its purpose and then take your audience through each stage of the process step by step. So our oral presentation on migration might go something like this:

> What is GPS? GPS stands for Global Positioning System and, as its name suggests, it forms a worldwide navigation system meaning you should never get lost again! As we can see on the second slide, the system is based on a network of satellites linked to ground stations around the world. To find where you are, the system measures the distances between your personal unit and three or four satellites, as you can see here . . . The exact location is then worked out by triangulation.

What you need to keep in mind is that you are trying to communicate information in manageable chunks, helping the audience to understand the progression of your argument or process. It doesn't matter at this early stage if you can't sequence your main points precisely (for instance, will it be better to deal with Wildebeest or Canada geese first?) because such a level of precision is not necessary at this stage. The important thing is simply to choose what your main points will be and make a good attempt at the sequencing; you can always revise it later on.

9.5.2 Select supporting information

We have already identified that there is a temptation to cram too much information into a presentation, simply because you have found lots of information at the research stage. An important arbiter that we have referred to twice previously in this chapter is the relevance of the material (to both the brief and the audience). However, there are other criteria that are helpful to consider explicitly when you get to the stage of selecting supporting information. The other criteria that we will consider are clarity, authority, and interest.

Clarity

When deciding whether information should be left in or taken out, ask yourself this question: 'Is this information that I'm thinking of including going to make what I am trying to say any clearer?' Of course, this relies on you knowing what it is you are trying to say, which is why establishing the main points first is important. For instance, in your research you may have found out lots of information about the global positioning system, including that the first experimental satellite for GPS was launched in 1978; but will this piece of information help to clarify what you are trying to say? If you have decided that the focus of your presentation is the examples themselves (Wildebeest and Canada geese) then you will probably decide that incidental detail about the history of GPS is unnecessary.

Authority

Ask yourself: 'Is this information that I'm thinking of including going to make what I am trying to say any more authoritative?' The authoritativeness of information you use will depend largely on the source of that information; whether or not it is well researched, impartial, and up to date. This is usually easier to establish for publications such as text-books and articles published in research journals than for web resources, especially freely editable sources such as Wikipedia. The authoritativeness of sources is addressed in Chapter 3, *Working with different information sources*.

Interest

Finally: interest. Ask yourself this question; 'Is this information that I'm thinking of including going to make what I am trying to say any more interesting?' Perhaps you have found a picture, chart, or even a video clip, which will help illustrate a point. Perhaps you could bring in a hand-held GPS device to illustrate how compact they are and how easily the technology can be applied to the study of migrating populations. Consider anything that will bring what you are saying to life.

So, establishing whether information you have found will make your presentation clearer, more authoritative, or more interesting is a helpful means of deciding whether or not it should be included. But always come back to the bigger, more fundamental question, 'is this relevant?' Also remember that when you decide to leave information out it doesn't necessarily mean that the information is wasted: the information you leave out may have been helpful in leading you to other information. It may also be useful supplementary information if you are asked questions about your presentation, or it may be useful if you subsequently have to write an associated essay or report.

9.6 **Prepare your notes**

Once you have analysed the brief, researched the topic, and planned and ordered your material, you are ready to prepare your notes. You have two main options available to you here: you could write them out word-for-word (known as *verbatim*) or you could use outline notes (e.g. bullet points). Alternatively, you could use a combination of these two types. We will consider each of these two options in turn and then look at how they can be combined to best effect.

9.6.1 **Verbatim notes**

Verbatim notes have many advantages and disadvantages.

Advantages of verbatim notes

Verbatim notes have the obvious advantage that you start the presentation knowing exactly what you are going to say. Knowing exactly what you are going to say can be reassuring and so boost your confidence. Verbatim notes can be particularly helpful, therefore, if you are lacking confidence in a particular presentation:

- perhaps because your first language isn't English;
- or you are presenting information that you are relatively unfamiliar with;
- or you are presenting complex material that requires a level of precision that would be difficult to achieve from outline notes;
- or you have some other reason for feeling nervous.

Disadvantages of verbatim notes

Verbatim notes do have significant disadvantages. There is a danger that you will simply read out your presentation and therefore won't communicate it effectively. When you look up towards the audience you might lose your place in your notes. Also, a read-out presentation is difficult to bring 'alive' because:

- you will lose eye-contact with your audience;
- your voice will probably become more monotonous;
- you will have much less energy and enthusiasm about the delivery;
- as a result your presentation will be less engaging.

How to use verbatim notes

Verbatim notes seem to have got themselves a bad name. Often people will tell you not to use them because of the very real disadvantages listed above, but this advice usually comes from people who have become experienced presenters and so have forgotten about much of the anxiety that is often associated with presenting. However, as can be seen from the advantages, there are very real benefits too. The trick is to use verbatim notes in the right way. We will take each of the potential disadvantages in turn and suggest a strategy to help you avoid that particular pitfall.

There is a potential danger that you will simply read out your presentation
Don't! Reading out the presentation will nearly always result in the problems listed. You don't need to memorize your notes but you do need to have practised them sufficiently so that you can work from them as a prompt, as opposed to as a script.

If you look up from your notes you might lose your place
Quite possibly. So make sure you format your notes in such a way so as to help you not lose your place. For instance, increase the font size, increase the line spacing, use paragraph breaks, use bullet points, use annotations and highlight key words. Figure 9.3 shows verbatim notes that would be very easy for you to lose you place with; Figure 9.4 shows more helpfully formatted verbatim notes that dramatically reduce the chances of you losing your place.

FIGURE 9.3 Unhelpfully formatted verbatim notes.

The use of GPS in the study of migrating populations
Hi, my name's Sam and my presentation is about how GPS helps scientists to study migrating populations.

This presentation is divided into three main sections, as shown on the first slide. Firstly, I'll give a brief explanation of what GPS is. Then I'll explain how the technology has been applied to the study of migrating populations. Finally, I'll tell you about two specific examples: Wildebeest and Canada geese.

Firstly then, an explanation of what is GPS? GPS stands for Global Positioning System and, as its name suggests, it forms a worldwide navigation system meaning you should never get lost again! As we can see on the second slide, the system is based on a network of satellites linked to ground stations around the world. To find where you are, the system measures the distances between your personal unit and three or four satellites, as you can see here ... The exact location is then worked out by triangulation (see, you always knew that trigonometry you slogged through at school would come in handy).

The first GPS systems were developed in the late 1970s for the United States military. Since then the technology has developed considerably and in the 1980s became available for civilian use. GPS is now used for many civilian purposes, including: map making, land surveying, and, of course, navigation. The particular use we are interested in though is in its application to the study of migrating populations of animals. Studying migrating populations using GPS has only been possible since the technology required to track animals has reduced in size. The necessary tracking tags are now small enough to track many different sorts of animal, not just large ones. Tags then relay information, via the GPS satellite system, to monitoring units, thus allowing scientists to know the precise coordinates of the animals they are studying. Such detailed information is invaluable for scientists as previously they were simply not able to know about an animal's whereabouts in such detail.

FIGURE 9.4 Helpfully formatted verbatim notes.

The use of GPS in the study of migrating populations
Don't start until you're ready
Introduction
Hi, my name's Sam and my presentation is about how GPS helps scientists to study migrating populations.

This presentation is divided into three main sections, as shown on the <u>first slide</u>.

1. Firstly, I'll give a brief explanation of what GPS is.
2. Then I'll explain how the technology has been applied to the study of migrating populations.
3. Finally, I'll tell you about two specific examples: Wildebeest and Canada geese.

Outline slide

1 min. max.

What is GPS?
Firstly then, an explanation of what is GPS? GPS stands for Global Positioning System and, as its name suggests, it forms a worldwide navigation system meaning you should never get lost again!

As we can see on the <u>second slide</u>, the system is based on a network of satellites linked to ground stations around the world. To find where you are, the system measures the distances between your personal unit and three or four satellites, as you can see here ...

A read-out presentation is difficult to bring 'alive' because you will lose eye-contact with your audience, your voice will probably become more monotonous, you will have much less energy and enthusiasm about the delivery, and as a result your presentation will be less engaging.

Again, most of the above will be helped by practice. If you can get to the stage where you have practised your presentation sufficiently so that your notes are a prompt rather than a script, you should be able to think about the communication, as well as content issues—issues such as making eye-contact with the audience, varying the tone of your voice appropriately, and generally trying to be more engaging.

9.6.2 Outline notes

Outline notes also have advantages and disadvantages.

Advantages of outline notes

Having your notes in only outline form can free you up to focus on engaging with the audience, rather than focusing on your notes. Also, if you are not reading your notes out, your voice and manner will probably be more natural, which will also help engage your audience. There is no temptation to read your notes out verbatim, which, as we have highlighted above, causes problems.

Disadvantages of outline notes

There are two main disadvantages of outline notes. Firstly, you might be more nervous about your presentation because you have fewer notes to rely on. Secondly, you may find that your mind 'goes blank' during your presentation because your outline notes aren't a sufficient prompt to remind you of what you intended to say.

How to use outline notes

As we have seen with verbatim notes, outline notes have both advantages and disadvantages; the trick is to use outline notes in the right way. The only way around these potential disadvantages is practice: you need to rehearse your presentation a sufficient number of times so that you know your outline notes are full enough to remind you of what you need to say. Of course, while you are testing the adequacy of your notes in this way you will also be learning the material. Remember though, just because you can get your presentation right once or twice in the privacy of your bedroom doesn't mean that everything will be fine in the more pressured situation of a tutorial class or assessed presentation.

9.6.3 Combine note types to maximize benefits

Given the potential disadvantages of each notes technique, probably the best method to employ is a mixture of the two. Where you are confident that you only need short bullet points to prompt you, you can use outline notes, for instance, for the introduction, as shown in Figure 9.5. However, where the material is more complex and precision is more important, you can use at least fuller, if not verbatim, notes, as shown in Figure 9.4.

FIGURE 9.5 Outline notes for introduction.

9.6.4 Openers and closers

In addition to planning the introduction, main points, and conclusion of your presentation, it is also important to think about how you will open the presentation and how you will close it. It is a good idea, regardless of what type of notes you choose to use, to know exactly what your opening statement is going to be (the 'opener') and exactly what your closing statement is going to be (the 'closer'). If you have a faltering start, perhaps tripping over your opening words or not saying them clearly, it is likely that you will lose confidence and composure, which can be very difficult to regain. Opening statements need only be simple sentences such as:

> 'Hi, my name's Sam and my presentation is about how GPS helps scientists to study migrating populations.'
>
> 'Hi, my name is Alex and I've got five minutes to talk to you about . . .'

Simple statements such as these get you off to a positive start and make it clear to the audience what you are doing. Equally, closing statements are important too. Too often presentations finish with a mumbled apology: 'Err, that's it, I've finished . . .', then the presenter sits down. A much better way to finish is a clear conclusion followed by a closing statement such as:

> 'So, in this presentation we have seen how GPS can be used as a very effective tool for monitoring the migration of very different types of animal, but also that there are some specific limitations that need to be addressed.
>
> Thank you very much for your attention and I will be happy to try to answer any questions.

9.6.5 **A note about working from visual aids**

An alternative to speaking from either verbatim or outline notes is to work from a visual aid. This is when you use a visual aid (e.g. a PowerPoint slide or an object) to prompt what you are going to say, rather than your notes themselves. This can work well because it frees you from your notes entirely yet can still give you sufficient prompts to remind you of what you want to say. However, as with verbatim and outline notes, it all depends on how the method is used. The method will work badly if your slides are essentially copies of your verbatim notes (as shown in Figure 9.6) and you end up reading them out. However, if you either condense the bullet points to key words or phrases, or better, use an image to illustrate what you want to say (as shown in Figure 9.7), you will find it much easier to talk from (this will be covered in more detail in Chapter 10).

FIGURE 9.6 A difficult-to-work-from slide.

What is GPS?
- The global positioning system (GPS) is the only fully operational global navigation satellite system
- More than 25 GPS satellites are in orbit around the Earth, transmitting signals that are picked up by GPS receivers, which, in turn, establish the receiver's location, speed and direction
- GPS is a crucial aid to navigation

FIGURE 9.7 An easier-to-work-from slide.

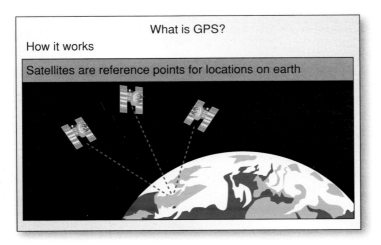

9.7 **Practice**

Finally, in this section on preparing for presentations we need to highlight the importance of practice. How effective a presentation is will depend to a large extent on how
much you have practised it. Practising a presentation is important because if you know
your presentation well it frees you up to think about the communication, as well as
content issues (style as well as substance).

   We noted above in *How to use outline notes* that just because you can get your presentation right once or twice in the privacy of your bedroom, it doesn't mean that everything will
be fine in front of an audience. However, the value of practising in your own environment
is not to be underestimated, and there are several ways you can make it more effective.

9.7.1 **Practise against the clock**

It can be very difficult to judge how much time your presentation will take without
actually practising it out loud and against the clock. Depending on the style of notes
you have used, sometimes you will look at your notes and think that it will barely last
2 minutes, never mind 5. Other times you will think you have got far too much material
but in reality you will get through it far quicker than you thought. The only way to have
a reasonably accurate idea of how long your presentation will take is to practise it out
loud. Of course, by doing this you are not only getting an idea of how long it takes, you
are also learning the material and refining its delivery.

9.7.2 Practise in front of a mirror

Just as it is difficult to know how long the presentation will take, it is also difficult to know how you will look. The only way to get an idea of this is to practise the presentation in front of a mirror or, better still, a video camera. You will probably feel a bit self-conscious doing this but it will give you some very useful feedback about how you are presenting yourself as well as how you are presenting your material. It is a particularly helpful way of assessing whether or not you are using your body language appropriately (see section 10.3.3 in Chapter 10).

9.7.3 Practise in front of your friends

Most usefully, you could practise in front of some willing friends. This could give you feedback on your presentation's length, your body language, the structure, in fact all aspects of your presentation. It can be a bit embarrassing and you will doubtless make mistakes, but as long as your friends are going to give some constructive comments, and you can deal with the potential embarrassment, it will give you invaluable feedback and so improve your final presentation, and thereby its grade, significantly.

✳ Chapter summary

The ability to present information verbally to groups of people is an increasingly important skill. It is of particular relevance to scientists as it is an important means of communicating research. This chapter began by identifying the characteristics of effective presentations and then identified the stages necessary to prepare presentations. Analysing the question or brief, researching the topic, planning and ordering the material, selecting supporting information, and preparing your notes were all identified as important aspects. Finally, the chapter addressed the issue of practising your presentation to ensure that you deliver it as well as you possibly can. This brings us on to the next chapter, *Delivering scientific presentations*.

Chapter 10

Delivering scientific presentations

➲ Introduction

You may be wondering why we have devoted an entire chapter just to the delivery of scientific presentations. Chapter 9 dealt in some detail with the preparation aspects of scientific presentations, so is a whole chapter purely on the delivery aspects really necessary? We think it is. As we identified in Chapter 9, *Preparing scientific presentations*, presentations can be divided into two overlapping but nonetheless distinct areas: content, and the communication of the content (or 'substance' and 'style'). We said that both of these elements are important and you need to be careful not to focus on one at the expense of the other. Just because you have prepared a good presentation (the content) it doesn't mean that you will be able to deliver it in a competent fashion (the communication of the content). Communicating the content involves a very different set of skills to preparing the content, so this chapter will look at how you can communicate your content to best effect.

In this chapter we will begin by highlighting common concerns people have about presentations and how to address them, then we will think about how to use visual aids effectively, and finally some key techniques to help your presentations run smoothly. But first, let's think about some important differences between written and spoken presentations.

Written versus spoken presentations

A useful starting point when considering the delivery of a presentation that you have prepared is to think about the similarities and differences between written and spoken presentations. One of the reasons some students get anxious about giving spoken presentations is that they consider them to be very different from written presentations. Of course, in many respects, the two forms are different, and we will come to the differences shortly, but there are many similarities too. In terms of the process that you need to go through in order to create written or spoken presentations, both forms require you to:

- analyse the question or brief;
- research the topic;
- plan and order the material;

- write the text or notes;
- review and re-draft the text or notes.

In addition to the above, in an academic context, both forms require you to present and analyse evidence, structure your argument, and reference your sources. When it comes actually to delivering the presentation, however, the major difference between the two forms is that for a written presentation all that is required is for you to hand it in (whether that be on paper or via email or web submission), whereas delivering a spoken presentation requires a lot more effort.

There are other important differences between written and spoken presentations. For example, when you are reading a textbook or journal article and you come across a complex argument that you don't understand you can skip back and re-read the argument until you do understand it. Or if you come across a large quantity of data that are difficult to take in, you can study the data more closely, or even look up additional information in other sources, to make them easier to understand. When taking in information in the context of a spoken presentation, however, it is often much more difficult to assimilate complex arguments or large quantities of information simply because you do not have this ability to skip back as you would with written material. Equally, if, when you are reading, you find yourself getting tired and your attention is wavering, you can simply stop reading, have a break, and come back to it later when you are feeling more alert. Clearly, when listening to a presentation, apart from walking out and choosing to miss the rest of the presentation (which would be a bit rude), you don't have this option. So in a presentation, presenters have to work hard to make sure their listeners stay with them, in terms of both their understanding and their attention. This chapter will explain how you can do this.

10.1 Common concerns and how to address them

The differences between written and spoken presentations are often the cause of concern or anxiety about delivering a presentation. For example:

- what if I'm so nervous I forget what I was going to say?
- what if the technology doesn't work?
- what if I lose my place in my notes?
- what if the audience doesn't seem to be following what I'm saying?
- what if I get asked questions that I don't know the answers to?

These, and other concerns, are important things to consider. Whilst listing possible concerns in this way can seem a little overwhelming, it is important to identify them so that something can be done about them. You might be able to relate to all of the above concerns, or perhaps to just a few of them, but it is likely that you will experience all of them at some point or other, so it will be useful to address each of them in turn.

10.1.1 **Deal with anxiety**

It's all very well saying 'deal with anxiety'; it is much more difficult actually to do it. It is also quite difficult to write about how to deal with anxiety, because different things make different people anxious. However, for the purposes of this chapter we are going to assume that standing up in front of a group of people to give a presentation is something that causes most people to be anxious or nervous. If it doesn't, it probably should!

We have deliberately titled this section 'deal with anxiety', rather than, for example, 'get rid of anxiety' for a couple of reasons. Firstly, you can't get rid of all your anxiety so it would be unrealistic to try; and secondly, anxiety or nervousness can be helpful, so, even if you could, you don't want to get rid of it altogether. Anxiety can motivate you to do the work needed so that you are effectively prepared for the presentation and it can also make you more alert and energized during the presentation. However, if your anxiety becomes too pronounced, difficulties can occur, which may impair your ability to prepare effectively for, and perform during, the presentation.

A useful strategy in dealing with anxiety is to identify what your concerns are so that you can address them, which is what this section is about. In general terms, however, the following strategies can be helpful:

- try to replace negative thoughts with positive ones (see Table 10.1);
- practise steady, deep breathing before and during your presentation;
- have a bottle of water to hand—it will help stop your throat drying out, but, better still, if your mind goes blank you can simply stop and take a sip, which usually gives you just enough time to gather your thoughts and regain your composure without the audience noticing.

TABLE 10.1 Replacing negative thoughts with positive ones

Negative thought	Positive replacement
'It will be a disaster.'	'I will aim to do the best I can.'
'I never do any good at this kind of thing, it's bound to go horribly wrong.'	'Just because I had a problem with this is in the past does not mean that things are bound to go wrong.'
'I will fail my degree and never get the career of my choice if I don't do well in this presentation.'	'The marks for this presentation are only a small percentage of my overall degree. If I don't do as well as I would like there will be other opportunities to improve my marks.'

10.1.2 **Have a back-up plan**

The second concern that we identified above was 'what if the technology doesn't work?' This is a legitimate concern and one that requires consideration. Broadly speaking there are two possible causes of technology apparently not working: firstly, it could be a genuine fault with the technology; or, secondly (and perhaps more commonly), it could be a lack of competence on the part of the person operating it. Both types of problem can be minimized by having a back-up plan, but be careful not to jump too quickly to using your back-up plan, when it could be, for instance, that you are just pressing the wrong button!

It is important, therefore, that you familiarize yourself with any technology that you intend to use. Amid the pressure and stress of a presentation about to start, or already started, it can be easy to panic and not think straight. Make sure you understand how to use what you are using. If possible, practise not just with something like it, but with the *actual* technology in the *actual* room where you will *actually* be presenting. The closer you can get to practising with the real thing the more confident and prepared you will be when it comes to the presentation.

For those occasions when the problem is with the technology rather than the person operating it, this is where you will have to use your back-up plan. Some forms of technology are less reliable than others, and the more multi-faceted the output the more potential there is for it to go wrong. For instance, if you are using PowerPoint, video, audio, and a live internet connection—there are a lot of things that can potentially go wrong! You need to consider whether or not such a range of technology is necessary to your presentation. There would be much less risk if you used less technology and it would be much simpler. Indeed, it may even improve your presentation (more technology does not necessarily mean better presentations).

10.1.3 Make your notes work for you

The third concern that we identified above was 'what if I lose my place in my notes?' As we identified in Chapter 9, there is a lot you can do to help yourself here. Firstly, make sure you use an appropriate type of notes: outline or verbatim. Secondly, make sure you format them in a way that makes them easy to use and so it is less likely that you will either lose your place (a possible problem with verbatim notes) or forget what you were going to say (a possible problem with outline notes). Chapter 9 also highlighted the benefits of combining these note types to maximize the benefits, using outline notes where you don't need much prompting and verbatim notes where you do.

10.1.4 Help the audience follow what you are saying

The fourth concern was 'what if the audience doesn't seem to be following what I'm saying?' This is clearly an important consideration for the benefit of the audience, but it is also a significant factor in your own performance as a presenter. If you begin to notice, during your presentation, that your audience is distracted and perhaps seemingly confused, it can be very difficult not to let that influence your confidence and you can begin to lose your composure as a result. Obviously, it is important not to be over-sensitive here; it can be easy to interpret normal audience behaviour, such as someone yawning, as a signal that your presentation is going really badly, when this probably is not the case. It is likely that there will always be some people in the audience who have had a bad night's sleep, or are thinking about what they are going to do at the weekend, or perhaps just exhibiting body language and facial expressions that are out of sync with their genuine response. Don't let one or two negative responses unduly influence your confidence. Instead, there are several things you can do during your presentation to help your audience follow what you're saying:

- use verbal signposts;
- use your voice;
- be aware of your audience's needs.

Let's consider each of these in turn.

Use verbal signposts

In Chapter 9 we identified that one of the characteristics of effective presentations was that they should be well structured. If a presentation is well structured then the audience can see the direction the presentation is taking and so should be able to follow it more easily. We said that one of the ways you can do this is by letting the audience know what the structure will be; both at the beginning and as you go along.

For example, at the beginning of a presentation you should map out where you are going by telling the audience what you will cover. The example presentation brief that we introduced in Chapter 9 was:

> Prepare and deliver a five-minute presentation on the use of global positioning systems (GPS) in the study of migrating populations.

For this presentation, letting the audience know what the structure will be may involve you saying something like this (as seen in Chapter 9): 'This presentation is divided into three main sections. Firstly I'll give a brief explanation of what GPS is. Then I'll explain how the technology has been applied to the study of migrating populations. Finally I'll tell you about two specific examples: wildebeest and Canada geese.'

In addition to mapping out the structure at the beginning of the presentation it is also helpful to give the audience pointers as you go along. If you were reading a written presentation you could simply flick through the pages and read the headings and subheadings, but clearly you can't do this with a spoken presentation. So help your audience follow what you are saying by telling them what the structure will be and then use pointers as you go along to make it clear where you are up to. The examples from Chapter 9 were:

- a single word such as 'Firstly', 'Secondly', and so on;
- linking statements such as 'So that's the background to the technology, let's now think about its application . . .';
- concluding statements such as 'In conclusion . . .' or 'Finally . . .'.

Use your voice

Use your voice to help your audience follow what you are saying. You can use it in many different ways by varying the volume, pace, and pitch.

Volume

Make sure that your voice is loud enough for your audience to hear clearly (it's surprising how many presenters don't do this—nervousness will often make you speak more quietly). Speaking too loudly or too quietly can make it difficult for your audience to follow your presentation. In normal conversation people tend to raise or lower their volume for emphasis. For example, a person may speak loudly when giving an instruction but softly when apologizing. Try to use these normal conversational variations in your own presentation.

Pace

Make sure that the speed of your delivery is easy to follow. If you speak too quickly (again, this is often caused by nervousness) or too slowly your audience will have difficulty following your talk. To add life to your presentation, try changing the pace of your delivery. A slightly faster section might convey enthusiasm. A slightly slower one might add emphasis or caution.

Pitch

The pitch of your voice also varies in day-to-day conversation and it is important to play on this when making a presentation. For example, your pitch will rise when asking a question; it will lower when you wish to sound severe.

Be aware of your audience's needs

If you are giving a 5- or 10-minute presentation to a group of your peers and a tutor, then you will be limited as to how much you can actually do about your audience's needs. You will probably be on quite a tight schedule (often you will lose marks for running over time) and if there are several of you presenting, then it will often be a case of just making sure the presentations fit in the allotted session. Additionally, when you are relatively new to presenting, you will want to concentrate on just delivering what you have prepared rather than making any changes to your presentation as you go along in response to the reaction you are receiving. However, if you are at least half way down a list of, say, six presentations that are scheduled for a session, there is one simple thing you can do (if your tutor doesn't suggest it): simply suggest a brief break. The break does not have to be long and it will probably be best if people don't even leave the room. However, if you suggest a break of a minute or so, while you are setting up, perhaps even suggesting that people stand up and walk around a bit, it should be just enough for people to wake up and re-energize themselves so that when you start your presentation they are that bit more attentive, making it easier for both you and them.

10.1.5 **Respond to questions appropriately**

The final concern we identified above was 'what if I get asked questions that I don't know the answers to?' This is a common concern because it is the bit of the presentation that you have least control over and it can be difficult to know what kind of questions you might be asked. If you are doing an assessed presentation, one important piece of information to find out is whether or not you are required to invite questions and if so whether or not your responses count towards the overall mark you receive. Whether or not the questions (or more accurately your answers!) count towards your mark, there are several things you can do to make this part easier.

Create a list of potential questions and prepare answers

We said above that it can be difficult to know what kind of questions you might be asked, but it is not impossible. In fact, with just a bit of thought you can probably come up with several potential questions that you might get asked. Questions will usually focus on areas of the presentation that the audience found particularly interesting, or perhaps a section that needed more explanation or additional context. Obviously you can't predict all the questions you might be asked, but you will be able to predict some, and that will help to take some of the pressure off.

Say whether you will be taking questions and when

It is a good idea to state clearly at the beginning of your presentation whether you will be taking questions and, if so, when. That way the audience is clear about if and when they can ask questions but also it means that you are more in control of the situation, which is important: lack of control, as identified above, can make the questions part of a presentation difficult. Sometimes people like to deal with questions as they go along as it can allow more interaction. However, it is usually easier to deal with questions at the end so you can deliver your talk free from interruptions, which you might find distracting.

Ask for clarification

When you get asked a question it is often unclear what the question actually means. If you are going to answer the question that the questioner wants to ask rather than the one you think they might be asking then it is important to ask for clarification. For instance, you might get asked the question:

> Do all the animals follow the same migrating pattern year after year?

A good way of clarifying the question is to rephrase it and then ask if that is what was meant. For example:

> When you say 'all the animals' do you mean you mean all migrating populations of wildebeest or all migrating populations of Canada geese?

Clarifying questions in this way is useful for a number of reasons:

- it gives the questioner chance to refine the question;
- it ensures that the rest of the audience has heard and understood the question;
- it enables you to check that you have understood the question;
- it also gives you thinking time to prepare a better answer.

Answer the question

Preparing potential questions and answers, saying if and when you will take questions, and asking for clarification will all help you to answer questions more appropriately. However, when you actually open your mouth to answer the question make sure you:

- answer the question you have been asked rather than the one you want to answer (seeking clarification helps here);
- are focused and to the point—answer the question and then stop, don't start waffling;
- are honest about the limitations of your knowledge—say if you don't know the answer (it will usually be obvious if you are bluffing);
- say if a question is beyond the scope of the presentation ('That's a good question but my research for this presentation focused on tracking migrating populations by GPS, not by any other means').

10.2 **Use visual aids effectively**

Using visual aids is relatively easy; it's using them appropriately that can be difficult. You will be limited to a certain extent by what is available, but it is equally important to consider what is appropriate; just because you could use something doesn't mean that you should use something. The list in section 10.2.1 shows the variety of visual aids available to you, and it is important that you recognize this variety; choosing the right form of visual aid is the first step in using them appropriately. Too often presenters default to using PowerPoint slides, without considering whether or not this is appropriate. For instance, a digital projector and laptop may be available to you, but it may not be appropriate to give a PowerPoint presentation if you are simply required to give an informal update regarding a project to a small tutorial group. In this situation, a concise handout identifying key points that you can chat through might be more suitable.

10.2.1 **Different types of visual aids**

The more common types of visual aid are as follows:

- whiteboard: large white board, usually fixed to the wall, can be written/drawn on with dry-wipe marker pens, either pre-prepared or spontaneously produced;
- flip chart: large pad of paper (A0), attached to a portable stand, can be written on with marker pens, either pre-prepared or spontaneously produced;

- handouts: usually A4 paper, handed out at an appropriate point (or points) during the presentation;
- objects: could be any physical object relevant to your presentation, shown from the front or handed round to your audience;
- video: shown on a television or via a laptop; various formats: VHS, DVD, or digital media file;
- overhead projector slides: hand-written or printed onto transparencies, either pre-prepared or spontaneously produced;
- Microsoft PowerPoint slides (or equivalent): produced on a computer and projected onto a screen via a digital projector; can be static or animated.

Each type of visual aid has advantages and disadvantages. These are summarized in Table 10.2.

TABLE 10.2 Advantages and disadvantages of various visual aids

	Advantages	Disadvantages	Suggestions
Whiteboard	<ul><li>They are usually big—so lots of space to write/draw</li><li>They are particularly good for drawing processes or equations because there is lots of space</li><li>Good for things you want to leave up and refer to at different points during your presentation</li><li>Good for recording comments or questions from the audience as you go along</li><li>Can be pre-prepared or spontaneous</li></ul>	<ul><li>If you write on it during your presentation you will be turned away from the audience as you are writing, which limits communication</li><li>If you run out of space you will need to rub off at least some of what you've previously written</li><li>You need to be able to write legibly and large enough for the audience to see</li></ul>	<ul><li>There aren't always appropriate pens available—so take your own</li><li>Write large and clearly in dark colours (black or blue)</li></ul>
Flip chart	<ul><li>Good for things you want to leave up and refer to at different points during your presentation</li><li>Good for recording comments or questions from the audience as you go along</li><li>Sheets can be torn off and stuck on the wall for reference —or flip back through the pad</li><li>Can be pre-prepared or spontaneous</li></ul>	<ul><li>Usually smaller than a white board so not suitable for larger groups</li><li>If you write on it during your presentation you will be turned away from the audience as you are writing, which limits communication</li><li>You need to be able to write legibly and large enough for the audience to see</li></ul>	<ul><li>There aren't always appropriate pens available or enough sheets—so take your own</li><li>Write large and clearly</li><li>Try to keep to one topic per sheet</li></ul>

TABLE 10.2 Cont'd

	Advantages	Disadvantages	Suggestions
Handouts	• The audience has something to take away for reference • Particularly good for supplementary information, e.g. references	• If handed out at the wrong time they can be distracting because the audience stops listening to the presenter and starts reading the handout • If handouts are very detailed it is more likely that the audience will listen less	• Use them for information that is too detailed to put on a slide • Think carefully about when and how to give them out
Objects	• Not a commonly used form of visual aid, so can have added impact • Other visual aids are mostly just looked at, objects give the audience something to touch too, therefore more tangible and memorable	• If you have only one copy of the object it will take time to hand round (so maybe better to just show from the front) • Because of the interest they can generate it is important that the object is central to your presentation, otherwise it will just be a distraction	• Make sure the object is directly relevant to the content of your presentation
Video	• Can add interest and help grab attention	• Can be hard to find a directly relevant video • The technology can be a little unpredictable	• Make sure you keep the clip short • Check the technology works
Overhead projector slide	• Projected image is usually large enough for even large groups to see • Can be hand-written or printed • Can be good for showing the development of a process or building up a diagram	• You need to be able to write legibly and large enough for the audience to see • Transparencies are expensive	• Make sure text is large enough to be seen from the back of the room • Keep slides simple, clear, and concise
PowerPoint slides	• Can be pre-prepared or spontaneous • Quick and easy to produce high-quality slides • Animation features can be helpful	• Some of the pre-set designs are unhelpfully formatted in terms of colour and background • Animation features can be very distracting if used inappropriately	• Keep slides simple, clear, and concise • Use animation only when appropriate

10.2.2 **When to use visual aids**

Visual aids can be used throughout your presentation, but too many visual aids can become distracting. Try to restrict the use of visual aids, therefore, to key parts of your presentation. Some suggestions of when to use visual aids are given below (we will address what visual aid to use shortly).

Introduction

You could use visual aids during the introduction to do some of the following:

- display the title of your presentation and perhaps your name;
- define particular terms or units you are going to use that your audience might not be familiar with;
- indicate what the structure of your presentation will be;
- display an image which introduces your subject or theme;
- highlight a question that you intend to answer during the course of your presentation.

Main points

You could use visual aids during the main body of your presentation to do some of the following;

- highlight your main points as they arise, using appropriate words or images;
- indicate the transition between main points;
- summarize your results or data visually using graphs, tables, maps, etc.;
- display key evidence from your research to support your argument.

Conclusion

You could use visual aids during the conclusion to do some of the following:

- summarize your main points;
- present your conclusion in a succinct phrase or image;
- list your references to enable your audience to read more on the topic.

10.2.3 **What visual aids to use**

In Chapter 9 we suggested a number of questions to help you select supporting information for a presentation. You can use these same questions to help you decide whether or not to use a particular visual aid. The fundamental issue is whether the material represented in the visual aid is relevant to the presentation, and if it is you then need to ask yourself; will this visual aid make my presentation:

- clearer?
- more authoritative?
- more interesting?

If your visual aid doesn't do any of the above three things, don't use it.

Visual aid choice exercise

For our sample presentation on global positioning systems and migrating populations a possible outline for the presentation was: 'Short explanation of what GPS is, how the technology has been applied to study migrating populations, and examples of wildebeest and Canada geese.' Imagine you were considering using the following visual aids for the above presentation:

- a flip chart page outlining the structure of your presentation (Figure 10.1);
- a handout giving supplementary details on how GPS works;
- pieces of card cut to 12 mm × 15 mm to illustrate the size of a GPS receiver;
- a video clip (2 minutes 41 seconds) of migrating wildebeest streamed from the National Geographic website (Figure 10.2);
- a hand-drawn overhead projector slide showing the migration path of Canada geese, as determined by GPS tracking studies;
- an image of a Canada goose on a PowerPoint slide (Figure 10.3);

Based on the questions listed above (section 10.2.3), decide whether or not you should include these visual aids in your presentation, and why.

Visual aid choice exercise feedback

We identified in section 10.2.3 above that the fundamental issue when choosing whether or not to use a visual aid is its relevance: is the material represented in the visual aid relevant to the presentation? If you decide it *is* relevant you then need to question whether the visual aid makes your presentation clearer or more authoritative or more interesting. We will deal which each of the suggested visual aids in turn, testing them against each of these criteria.

A flip chart page outlining the structure of your presentation (Figure 10.1)

This is clearly relevant to your presentation because it summarizes its content. Although it won't make the presentation more authoritative or more interesting, it could help make it clearer. If you were to use a flip chart in conjunction with another form of visual aid, for example an overhead projector, it would have the added advantage of the outline always being there for you to refer to. Flip charts are probably best used only when the audience is relatively small and close by.

FIGURE 10.1 Flip chart summarizing presentation content.

```
OUTLINE
..........
•   What is GPS?
•   Applications to migrating populations
•   Examples
    –   Wildebeest
    –   Canada geese
•   Conclusion
```

A handout giving supplementary details on how GPS works

This is potentially relevant but there is a danger of missing the main focus of the presentation. The presentation is primarily about the application of GPS in the study of migrating populations, it is not a presentation on GPS itself. A brief explanation of what GPS is and how it works is useful background information, but it only features in the presentation so that your audience can understand the main focus: its application to your two chosen migrating populations. So, basic information about GPS will help make your presentation clearer but it is only there to provide context. It may add greater authority to the presentation, but this depends largely on the source of the information (see Chapter 3, *Working with different information sources*). It could add interest, but this depends on how you communicate it (because of its potential lack of relevance it could actually make the presentation less interesting). However, if you genuinely think that the information is useful, a handout is probably the best format for it. Handouts are good for communicating (among other things) supplementary information; your audience can just refer to the handout if they are interested. Be careful about when you give them out though: you don't want your audience having their heads down reading when you are trying to move on to the next point!

Pieces of card cut to 12 mm × 15 mm to illustrate the size of a GPS receiver

This might sound a little bit unconventional, but that doesn't necessarily make it a bad idea. Again, it shouldn't be the main focus of the presentation, but it would illustrate well how the technology can be applied to the context in question. One of the reasons GPS can be used to track animals is because the tags required can be so small. Just saying that they are '12 mm × 15 mm' doesn't have much impact; showing with a visual aid how small they can be is much more striking. Better than saying '12 mm × 15 mm' would be to say 'they are about the size of your thumb nail', but actually handing out an appropriately sized piece of card adds an extra dimension, giving people something to hold and touch and think about. So using such a visual aid would make the point more clearly; it wouldn't add any authority but it might add some interest.

A video clip (2 minutes 41 seconds) of migrating wildebeest streamed from the National Geographic website (Figure 10.2)

This might sound exciting but it is probably inappropriate for a number of reasons. It is potentially relevant in terms of the brief because it is about the migrating habits of one of your chosen populations, but it's probably not relevant in terms of the level. In Chapter 9, we said that a presentation should be relevant not only with regard to the brief, but also with regard to the audience. A video designed for a popular audience is probably not going to be relevant for academic purposes. It probably won't make your presentation any clearer, unless it specifically addresses the use of GPS technology. It's unlikely to add authority because, even though it is from a credible source, the level, as identified above, is too popular. It might add interest but the clip is far too long; if you were to use the entire clip it would take up more than half your allotted time for your presentation.

FIGURE 10.2 Video of wildebeest migration.

A hand-drawn overhead projector slide showing the migration path of Canada geese, as determined by GPS tracking studies

If you like the idea of using cutting-edge technology you might be reluctant to use a hand-drawn overhead projector slide. However, let's think about our criteria. Such a slide would be relevant because it would be illustrating a very important part of your presentation. It would make the content clearer by representing visually the information that GPS technology has been used to gather on this particular migrating population. It adds authority because the GPS tracking studies are a primary source of information. It might even add interest because you are illustrating the vast distances these birds fly and the places to which they travel.

An image of a Canada goose on a PowerPoint slide (Figure 10.3)

The image is relevant in terms of the brief because it is a sample of one of your chosen populations. However, by itself it is not communicating very much information. It could be argued that it will help make your presentation clearer (because it makes it obvious which bit of your presentation you are on), that it doesn't add any authority but it could add some interest. It is appropriate to use images such as this, but make sure you have a clear purpose for using them (and as long as you acknowledge the source, see Chapter 8, *Avoiding plagiarism*). Don't just think, 'there's a picture of a Canada goose—I'll use that', ask yourself why you need to use it. Your audience probably knows what a Canada goose looks like, so that's not your reason, but your reason could be to use it as a background to the start of your section on Canada geese so that the audience are clear that you have moved on from wildebeest.

FIGURE 10.3 Image of a Canada goose in flight.

10.3 Master some key techniques

We have addressed some common concerns that people often have about giving presentations and have considered how to use visual aids appropriately. Finally, it's important to think about some key techniques that you need to develop if you are going to deliver your presentation effectively.

10.3.1 Relax!

The first key technique is to try to relax. Clearly, this is easy to say but much more difficult actually to do. However, if you can relax, even just a little bit, it will help your presentation a great deal. Presenters who are tense will deliver their presentation in a stilted, unnatural, and nervous manner, whereas a presenter who is relaxed (or slightly less tense at least) will be more natural and more confident in their delivery. We addressed many of the issues that will help you to be more relaxed in section 10.1.1, *Deal with anxiety*, but we also wanted to highlight it here as a key technique because it makes such a positive difference to a presentation.

10.3.2 **Be conversational**

In Chapter 9 we highlighted the importance of getting to the stage where you have practised your presentation sufficiently so that your notes (whatever their format) act as a prompt rather than as a script, thus freeing you up to think about the communication, as well as content issues. Something to aim for is to try to be conversational. Being conversational doesn't necessarily mean being informal or chatty (this is often not appropriate in an academic context) but it does mean that you are speaking in a natural way, as if you were actually having a conversation. Clearly, a 5-minute presentation is a fairly one-way conversation (!) but if you can aim to express yourself in normal, every-day (though still appropriate) language it will help the presentation to be more engaging, and being engaging was one of the key characteristics of effective presentations that we mentioned in Chapter 9.

10.3.3 **Think about your body language**

Body language is an important and often neglected element of a presentation. It includes facial expressions and eye-contact as well as what you do with your hands and where and how you situate yourself in the room. Body language can be both positive and negative: it can help a presentation be more engaging, or it can distract from and even counteract the presentation's content. If you have practised your presentation in front of your friends, a mirror, or a video camera (as suggested in Chapter 9) you will be more aware of how your body language influences the way you present. Even though it can be counterproductive to become too self-conscious of your body language, here are some suggestions to make sure you use your body language positively.

Think about how you will use the available space

Where are you going to stand in the room (you might prefer to sit but it's more normal to stand and it helps you project your voice better)? Where will you put your notes? If you are using PowerPoint slides or an overhead projector you need to be able to stand somewhere that doesn't obscure anyone's view of the screen but still allows you to be close enough to change slides without getting in anyone's way or tripping over cables. Spend some time thinking about this before you start presenting because once you have started your presentation you want to be able to concentrate on what you are saying rather than thinking about where the best place is for you to stand.

Think about what you will do with your hands

People use their arms and hands in everyday conversation to add emphasis or to help describe events. Presenters will therefore look rather awkward if they keep their hands in their pockets or rooted firmly at their sides. Equally, you don't want your hands to move in a way that distracts your audience's attention such as nervous, repetitive movements. Instead, use your hands to emphasize and enhance what you want to say verbally. As long as use of your hands is controlled and purposeful, then it has a key role to play in supporting your verbal communication. If you can try to relax a little and be conversational then it is more likely that what you do with your hands will be more natural and more in sync with what you are saying.

Make eye-contact

Often during an assessed presentation the presenter only makes eye-contact with the person doing the assessing, thereby excluding everybody else. If you were to have a conversation with someone you would make eye-contact with them, and not doing so would probably be considered a bit rude. So when presenting, try to make eye-contact with everyone. It is likely that you will only be presenting to a relatively small number of people, so making eye-contact with everyone is a realistic aim. If, however, you are presenting to larger groups just try to make eye-contact with all parts of the room. Making eye-contact can be more difficult if you are using verbatim notes, so make sure you follow the guidelines on how to use verbatim or outline notes in Chapter 9.

Smile

If you are nervous it can be difficult to smile, but smiling can make a big difference to how people react to you. Clearly, you don't want to have an inane grin on your face for the entire presentation—that would be unnatural and look ridiculous—but making a point of trying to smile, at least before you start, will help your audience feel more at ease. It is likely that at least someone in the audience might smile back at you, which in turn helps you feel more at ease too.

10.3.4 **Show some enthusiasm!**

Finally, show some enthusiasm! If you appear bored by and uninterested in what you are saying, your audience will probably react in the same manner. Conversely, if you can show genuine interest and enthusiasm it is more likely that your audience will be willing to listen attentively and work at following what you are saying.

....

✳ **Chapter summary**

There are similarities between written and spoken presentations, but there are significant differences too. It is often the differences that are the cause of concern or anxiety associated with delivering a presentation. This chapter has addressed these common concerns and has also focused on how to use a range of visual aids appropriately and the importance of mastering some key techniques. Following these guidelines will ensure that you deliver your scientific presentations much more effectively.

Chapter 11

Creating academic posters

⊃ Introduction

Poster presentations are a common academic format and are a regular feature at academic conferences. As an undergraduate, it is unlikely that you will be presenting a poster at a conference, but the poster format is sometimes used as a form of assessment for undergraduate courses. This chapter will explain the purpose of academic posters and suggest seven key steps for creating them effectively.

11.1 The context of poster presentations

If posters are used as a form of assessment at undergraduate level then they have a different purpose from that of an academic conference. However, it is helpful to understand the context from which they come in order to know what they are supposed to achieve.

Academic conferences are important opportunities for researchers to present their work, allowing them to:

- publicize their findings;
- promote discussion;
- create opportunities for collaboration with other researchers.

Clearly, not everyone at a conference will be able to present their research orally, so poster presentations are often used to allow lots of people to present their research at once. Typically, posters are size A1 (eight times bigger than A4) or A0 (16 times bigger than A4) and are attached to display boards arranged throughout an open plan room or conference hall. Time will be allocated in the conference schedule for delegates to view the posters, usually with opportunities to discuss the work with the presenter. A poster presentation is more limited than an oral presentation in terms of the amount of information that can be communicated, therefore posters should be designed to summarize the important elements and promote discussion.

Depending on how large the conference is and how well it is organized, delegates may have brief details of posters on display at the conference in advance, including the title of the poster and possibly an abstract. If these details are available then delegates might scan the list of posters on display to identify those they wish to view. Alternatively (or additionally), delegates may simply browse the posters themselves and view those which catch their eye. Either way, the posters that are viewed the most will be those that stand out from the crowd in some manner.

11.2 Seven key steps to creating your poster

Now that we have explained the context of academic posters we need to identify how to create them effectively. There are seven key steps involved in the process; the first two relate to the content of the poster, the next four are about design issues, and the final one is about ensuring that it is free from errors. This may seem a little unbalanced but we have dealt a lot already in this book with writing for bioscience, and it is usually the design aspects, rather than the content aspects, where students need the most guidance. The seven steps, therefore, are:

1. Get the academic content right.
2. Cut the text down to between 300 and 500 words.
3. Format the type.
4. Prepare your images.
5. Design the layout.
6. Choose a colour scheme.
7. Check it very carefully.

11.2.1 Get the academic content right

Firstly you need to make sure that you get the academic content right. This is the foundation for everything else that follows. If you don't do this first you will waste a lot of time formatting and laying out content that you end up not using. Usually, for undergraduate assessment purposes, you will be given a specific brief for your poster (often based on another aspect of assessment such as a practical report, essay topic, or research method). This brief will determine the parameters of the academic content. In addition to the academic content you may also be given guidance on the audience the poster is intended to be designed for. The intended audience will determine the level of the academic content that is appropriate.

Specialist

If the audience is from the same specialist field as the presenter, the presenter can assume that the people viewing the poster will have a high level of knowledge about the subject. It is therefore acceptable for the presenter to use technical language and terms on the poster.

Related

If the audience is from a field related to that of the presenter, the presenter can only assume that the people viewing the poster will be familiar with the generalities of the discipline. The presenter, therefore, needs to be careful about using technical language and terms.

General

If the audience is from the general public the presenter can only assume general knowledge. The people viewing the poster will probably have no familiarity with the discipline or subject area. The presenter, therefore, needs to avoid using technical language and terms and should use only basic descriptions.

For the purposes of this chapter we will use the following brief for a poster.

> Design and produce an academic poster summarizing a current research technique in biological sciences. The intended audience is first-year undergraduate bioscience students. The poster needs to be size A1 and landscape orientation.

The current research technique we will use is *animal cloning*. We will follow this example through in the remaining six steps of the process.

11.2.2 **Cut the text down to 300–500 words**

The second step in creating your poster is to use an appropriate number of words. Clearly, from the title of this section we are assuming that initially you will have too many words rather than too few. That's because a poster usually requires you to take a piece of writing you have already created for another purpose and summarize its content. At an academic conference the original writing will usually be some form of research; at an undergraduate level, as identified above, it will usually be based on another aspect of assessment such as a practical report, essay topic, or research method. The recommended number of words for a poster is between 300 and 500; this may not seem like much (and it isn't!) but it is important to keep the number of words within these limits in order to create a poster that is both readable from a distance and conveys key information quickly. The number of words affects readability because the greater the number of words the smaller the font size needs to be in order to accommodate them on the page, and so the less readable it becomes. Also, if the poster contains too much text, the amount of information can be overwhelming to the reader, making it less likely that passing delegates will spend time viewing it. To get an idea of what 300 and 500 words looks like, see Figures 11.1 and 11.2, respectively.

FIGURE 11.1 A poster with 300 words.

FIGURE 11.2 A poster with 500 words.

When reducing the text down to between 300 and 500 words you need to remember that you are trying to make it easy for a person who is not familiar with the content to understand it quickly. It can be helpful to try to think of it a bit like a trailer for a movie: you are not trying to tell people everything there is to know about the subject, rather you are just giving an overview in order to summarize the main points and encourage people to find out more.

For our poster on animal cloning we have summarized the current research techniques that are involved in the process and have tried to explain them appropriately for an audience of first-year undergraduate biological sciences students, as per the requirements of the brief. We have used a conventional scientific report pattern to provide a clear structure and have endeavoured to focus only on the essentials of animal cloning to create a succinct text. The words for our animal cloning poster are as follows:

Introduction

Animal cloning is a contentious topic. The history of cloning dates back to the 19th century (Driesch, 1892); however, animal cloning came to particular public prominence in 1997 when Wilmutt *et al.* demonstrated that the cloning of mammals from adult cells, rather than an embryo, was possible. The authors of the study derived nuclei of cultured mammary epithelial cells from an adult ewe, resulting in the birth of 'Dolly'—the first cloned sheep.

Methods

Cloning from adult cells involves taking the nucleus of a cell from the animal intended to be cloned and transferring it into an unfertilized ovum which has had its metaphase chromosomes removed. If this step is successful, the transferred nucleus is 're-programmed' to initiate the development of a new embryo. This embryo is genetically identical to the animal from which the cell was obtained. The embryo is then transferred into a surrogate mother for gestation to term and birth.

However, there are many factors affecting the success of cloning, including: the type of nuclear donor cell used; the treatment of donor cells prior to nuclear transfer; the source of recipient ova; the techniques employed; and the competency of the laboratory group working on it.

Cloning in cattle

There are more laboratories worldwide cloning cattle than all other species combined. This is partly due to the knowledge and expertise of assisted reproductive techniques in cattle developed over the last 20 years. Whilst the success and repeatability for cloning from embryonic cells are now well established in cattle, cloning from adult cells still remains more problematic.

Cloning in pigs

The cloning of pigs has illustrated well the fact that assisted reproductive technologies and techniques for nuclear transfer do not directly apply from one species to another. In a study by Onishi *et al*. (2000) the transfer of 110 cloned embryos to four surrogate mothers produced only one piglet.

Results

Table 1 shows the success rates of animal cloning from mammalian adult cells. The figures are taken from a review article by Solter (2000).

Species / Number of oocytes manipulated / Success rate (%)	Cow / 99 / 2
	Mouse / 1,345 / 0.7
Sheep / 277 / 0.4	Pig / 401 / 1.2

It is evident from the above figures that, even given the significant progress made in the field of animal cloning from adult cells in recent years, success rates are still very low.

Conclusion

Techniques used to clone animals are still developing, and cloning to reproduce certain genotypes can represent a significant challenge. With a variety of cloned animals being produced, the demand for cloning specific animals will increase significantly in the coming years. Differences in the ability to clone a particular species are a result of the time and resources which have been invested into research.

References

Driesch, H. 1892. Entwicklungsmechanische Studien. I. Der Wert der beiden ersten Furchungszellen in der Echinodermentwicklung. *Zeitschrift für Wissenschartliche Zoologie*, **53**, 160–178.

Onishi, A., Iwamoto, M., Akita, T., Mikawa, S., Takeda, K., Awata, T., Hanada, H., Perry, A. 2000. Pig cloning by microinjection of fetal fibroblast nuclei. *Science* **289**(5482), 1188–1190.

Solter, D. 2000. Mammalian cloning: advances and limitations. *Nature Reviews Genetics* **1**, 199–207.

Wilmut, I., Schnieke, A.E., McWhir, J., Kind, A.J., Campbell, K.H. 1997. Viable offspring derived from fetal and adult mammalian cells. *Nature*, **385**(6619), 810–813.

The above text contains 462 words (not including the references). The structure follows a conventional scientific report pattern: introduction, methods, results, conclusion, and references.

11.2.3 **Format the type**

Thirdly, you need to format the type. When formatting type for a poster, you need to remember two important principles: to use consistent styles, and to group sections of text appropriately.

Use consistent styles

Using consistent styles helps achieve what graphic designers call the principle of **resemblance**. It means that text which performs the same function should always be formatted in exactly the same way. For instance, if you decide that the subheadings should be Arial, 10 point, bold, left-hand justified with an after paragraph line spacing of 0.2 lines then all your subheadings should be formatted in exactly this way. Similarly, you might decide that the captions for figures should be Arial, 6 point, regular, right-hand justified with a before and after paragraph line spacing of 0.2 lines, in which case all your captions should be formatted in exactly this way.*

There are two reasons for using consistent styles: firstly it makes the poster look neater; secondly it makes the structure of the poster easier to understand because you know which text is performing what function. An example of using consistent styles is shown in Figure 11.3.

FIGURE 11.3 Using a limited number of consistent styles.

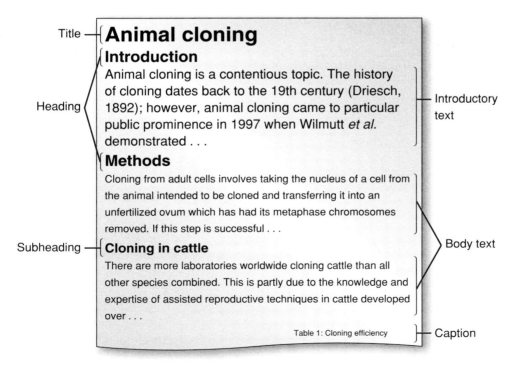

*Note that 10 point and 6 point are small font sizes, but posters are often created in A4 and then enlarged to A1 or A0, making the finished text eight or 16 times bigger respectively. See Box 11.1.

BOX 11.1 How do I know what font size to use?

Posters are often created at A4 size and then enlarged to the appropriate size for a poster, usually either A1 or A0. This can make it difficult to know what font size is appropriate to use because you are not creating your poster at the size of the finished product. The magnification factor from A4 to A1 is approximately 300% (it is actually 283%, for reasons that we won't go into here) and from A4 to A0 is 400%. This relationship is illustrated for you in Figure 11.4.

FIGURE 11.4 A4 paper to A1 or A0.

Therefore, a font size of 10 point at A4 is equivalent to a font size of 40 point at A0. Approximate appropriate font sizes to use for A1 and A0 posters are shown in Table 11.1.

TABLE 11.1 Suggested font sizes to allow appropriate magnification

Style	A4 to A1	A4 to A0
Title	**18 point Arial bold**	**16 point Arial bold**
Heading	**13 point Arial bold**	**12 point Arial bold**
Subheading	**11 point Arial bold**	**10 point Arial bold**
Introductory text	11 point Arial bold	10 point Arial regular
Body text	10 point Arial bold	8 point Arial regular
Captions	8 point Arial bold	6 point Arial regular

Group sections of text appropriately

Grouping sections of text appropriately helps achieve what graphic designers call the principle of **proximity**. It means that text which belongs together is grouped together; this is usually achieved using paragraph spacing. Figure 11.5 illustrates inappropriate spacing of text; there is a whole line space between each section and in the bulleted list. The result is that the headings and bullets float around and it is not immediately clear which piece of text belongs to which section.

FIGURE 11.5 Inappropriate proximity of text.

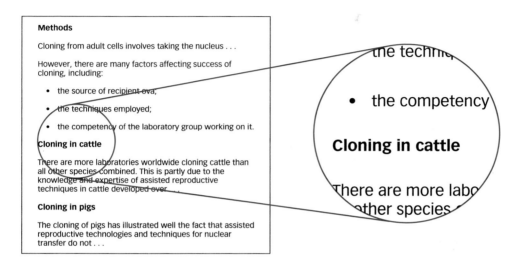

Figure 11.6 illustrates more appropriate spacing of text; paragraph spacing has been used to make it clear which pieces of text the headings and bullets belong to.

FIGURE 11.6 Appropriate proximity of text.

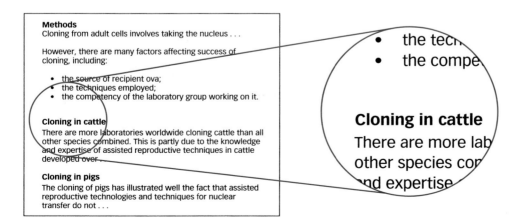

Line spacing, alignment, and text formatting

In addition to thinking about the principles of resemblance and proximity when formatting the type, it is also important to consider line spacing, alignment, and text formatting. Figure 11.7 illustrates the effect of line spacing on body text; body text (which is usually the smallest text on the page other than captions) is much easier to read if the line spacing is slightly increased.

FIGURE 11.7 Line spacing and body text.

Line spacing = 0.9
It is evident from the above figures that, even given the significant progress made in the field of animal cloning from adult cells in recent years, success rates are still very low.

Line spacing = 1.0
It is evident from the above figures that, even given the significant progress made in the field of animal cloning from adult cells in recent years, success rates are still very low.

Line spacing = 1.1
It is evident from the above figures that, even given the significant progress made in the field of animal cloning from adult cells in recent years, success rates are still very low.

Titles that run to more than one line, however, often benefit from the reverse: compressed line spacing, as shown in Figure 11.8. However, where possible, it is better if you can just make your titles more succinct so that they can fit on one line.

FIGURE 11.8 Line spacing and titles.

Large text (e.g. titles) probably needs decreased line spacing . . .

Line spacing = 0.9

. . . because it looks a bit strange with increased line spacing.

Line spacing = 1.1

It is recommended that you left-hand justify all body text (as shown in Figure 11.9) because fully justifying the body text results in awkward gaps on some lines, making the text more difficult to read (as shown in Figure 11.10).

FIGURE 11.9 Left-hand-justified body text.

Animal cloning is a contentious topic. The history of cloning dates back to the 19th century (Driesch, 1892); however, animal cloning came to particular public prominence in 1997 when Wilmutt *et al.* demonstrated that the cloning of mammals from adult cells, rather than an embryo, was possible. The authors of the study derived nuclei of cultured mammary epithelial cells from an adult ewe, resulting in the birth of 'Dolly'—the first cloned sheep.

FIGURE 11.10 Fully justified body text.

Animal cloning is a contentious topic. The history of cloning dates back to the 19th century(Driesch, 1892); however, animal cloning came to particular public prominence in 1997 whenWilmutt *et al*. demonstrated that the cloning of mammals from adult cells, rather than an embryo, was possible. The authors of the study derived nuclei of cultured mammary epithelial cells from an adult ewe, resulting in the birth of 'Dolly'—the first cloned sheep.

Finally on formatting the type:

- use the same font throughout your poster (unless using a complementary font for headings);
- set the headings in **bold**;
- use *italics*, *underlining*, and CAPITALS sparingly (only when convention dictates);
- break up any large areas of text with subheadings.

11.2.4 **Prepare your images**

The fourth step is to prepare your images. Images can be photographs, diagrams, graphs, or tables. The appropriate use of images can benefit a poster considerably, not just in terms of its visual impact, but also in terms of how readily understandable the content of the poster is. We will deal with each of these image types in turn.

Photographs

There are three key points you need to remember about any photographs you use on a poster.

1. They need to be relevant to the content of your poster, not just put in to brighten it up.
2. If they are not your own photographs they need to be appropriately referenced (in the same way as you would reference text that wasn't your own).
3. They need to be of an appropriate resolution.

The resolution is especially important if you are creating a poster in A4 format and then enlarging it to A1 or A0. In order for photographs in particular, and images generally, not to degenerate, they need to be of sufficient quality to start with—that is, they need to be of a sufficiently high resolution. The resolution of digital images is measured in pixels per square inch. Photographs copied from the internet usually have a resolution of 72 pixels per inch and so will look 'grainy' or 'pixelated' when enlarged, as shown in Figure 11.11.

For a printed poster, choose images which have a resolution of at least 300 pixels per inch at the size at which they will appear on the poster. In all likelihood you will want to enlarge the image when you reproduce it for the poster: and as soon as you

FIGURE 11.11 The effect of enlargement on low resolution images.

enlarge it, the resolution will decrease. For example, if you have an image that is 2 cm square and has a resolution of 300 dpi, when you enlarge it to become 4 cm square, its resolution will drop to 150 dpi (which is too low a resolution for printing purposes). Instead, start with an image that has a resolution that's well above 300 dpi. Then, when you enlarge it, the resolution will drop—but hopefully not below 300 dpi.

Diagrams

A diagram can be a very useful way of communicating information in a quick and succinct manner. Therefore, they are of particular use in posters, where communicating information in a readily understandable and concise fashion is very important. A well-chosen diagram can save a lot of words or make text descriptions much easier to understand. In our example poster, Figure 11.12 supplements the information in the first paragraph of the methods section to make it more readily understandable.

Graphs

When making graphs for posters, you need to think carefully about how you format them, as viewing them from a distance means that default formatting is usually inappropriate. The graph in Figure 11.13 is difficult to view from a distance because of the relatively small font, the shaded plot area and the excessive use of horizontal lines. All these formatting features are default features in Microsoft Excel. As a general rule, when it comes to the formatting of a graph, don't accept the default formatting provided by your software.

FIGURE 11.12 A diagram saving a lot of words.

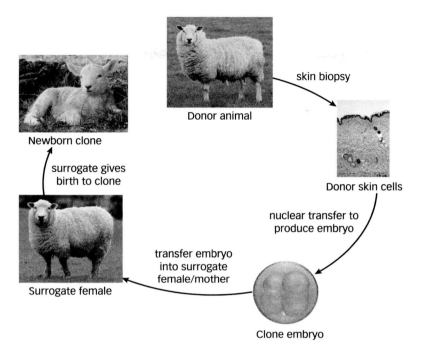

FIGURE 11.13 An inappropriately formatted graph.

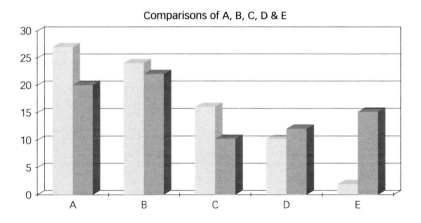

Figure 11.14 shows a more appropriately formatted graph. The font size has been increased, the shading on the plot area has been removed, and the use of horizontal lines has been minimized. Additionally, the colours of the bars have been altered to provide greater contrast; making the two sets of figures easier to distinguish between. Again, as with photographs and diagrams, if the data are not your own you need to reference them appropriately.

FIGURE 11.14 A more appropriately formatted graph.

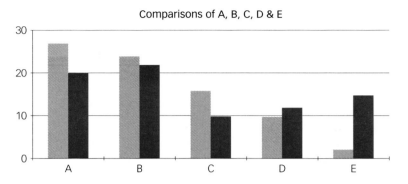

Tables

Finally in this section on preparing images, let's consider tables. The principles that applied to graphs apply equally to tables: format them in such a way so as to make them clear, simple, and easy to see. Again, default formatting is usually inappropriate as it tends to make the text too small and the grid lines can make patterns in the data difficult to discern. As a general rule, don't format tables that look like spreadsheets, as shown in Figure 11.15, as they are more difficult to interpret, even when you are dealing with a small data set.

FIGURE 11.15 Default formatted table.

Species	Number of oocytes manipulated	Success rate (%)
Sheep	277	0.4
Cow	99	2
Mouse	1345	0.7
Pig	401	1.2

Figure 11.16 shows a table that has been formatted more appropriately. The following changes have been made:

- the font size has been increased;
- the header row is made clear by setting the text in bold and lightly shading the row;
- the internal vertical lines have been deleted to allow patterns between the rows to be more easily seen;
- the numbers have all been right-hand justified so that the decimal place is vertically aligned;
- the percentages have all been quoted to the same number of decimal places;
- the rows have been ordered by the magnitude of the aspect of the data that attention needs to be drawn to: the percentage success rate.

FIGURE 11.16 A more thoughtfully formatted table.

Species	Number of oocytes manipulated	Success rate (%)
Cow	99	2.0
Pig	401	1.2
Mouse	1345	0.7
Sheep	277	0.4

11.2.5 Design the layout

The fifth step is to design the layout. The layout needs to be designed so that a person viewing your poster can quickly understand the sequence of the information. This is most easily achieved using appropriate and conventionally ordered headings, as shown in Figure 11.17.

FIGURE 11.17 Use of appropriate headings to show sequence.

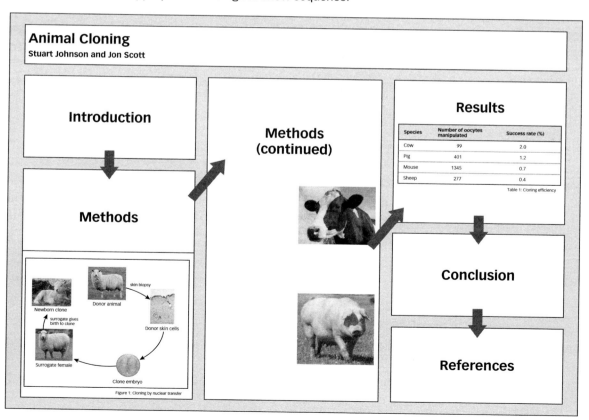

FIGURE 11.18 Use of a design grid to show sequence.

Animal Cloning
Stuart Johnson and Jon Scott

Introduction
Animal cloning is a contentious topic. The history of cloning dates back to the 19th century (Driesch, 1892); however, animal cloning came to particular public prominence in 1997 when Wilmutt *et al.* demonstrated that the cloning of mammals from adult cells, rather than an embryo, was possible. The authors of the study derived nuclei of cultured mammary epithelial cells from an adult ewe, resulting in the birth of 'Dolly'—the first cloned sheep.

Methods
Cloning from adult cells involves taking the nucleus of a cell from the animal intended to be cloned and transferring it into an unfertilized ovum which has had its metaphase chromosomes removed. If this step is successful, the transferred nucleus is 're-programmed' to initiate the development of a new embryo. This embryo is genetically identical to the animal from which the cell was obtained. The embryo is then transferred into a surrogate mother for gestation to term and birth (see Figure 1).

Figure 1: Cloning by nuclear transfer

However, there are many factors affecting success of cloning, including:

- the type of nuclear donor cell used;
- the treatment of donor cells prior to nuclear transfer;
- the source of recipient ova;
- the techniques employed;
- and the competency of the laboratory group working on it.

Cloning in cattle
There are more laboratories worldwide cloning cattle than all other species combined. This is partly due to the knowledge and expertise of assisted Reproductive techniques in cattle developed over the last 20 years. Whilst the success and repeatability for cloning from embryonic cells is now well established in cattle, cloning from adult cells still remains more problematic.

Cloning in pigs
The cloning of pigs has illustrated well the fact that assisted reproductive technologies and techniques for nuclear transfer do not directly apply from one species to another. In a study by Onishi *et al.* (2000) the transfer of 110 cloned embryos to four surrogate mothers produced only one piglet.

Results
Table 1 shows the success rates of animal cloning from mammalian adult cells. The figures are taken from a review article by Solter (2000).

Species	Number of oocytes manipulated	Success rate (%)
Cow	99	2.0
Pig	401	1.2
Mouse	1345	0.7
Sheep	277	0.4

Table 1: Cloning efficiency

It is evident from the above figures that, even given the significant progress made in the field of animal cloning from adult cells in recent years, success rates are still very low.

Conclusion
Techniques used to clone animals are still developing, and cloning to reproduce certain genotypes can represent a significant challenge. With a variety of cloned animals being produced, the demand for cloning specific animals will increase significantly in the coming years. Differences in the ability to clone a particular species are a result of the time and resources which have been invested into research.

References
- Driesch H 1892. Entwicklungsmechanische Studien. I. Der Wert der beiden ersten Furchungszellen in der Echinodermentwicklung. *Zeitschrift für wissenschaftliche Zoologie* 160–178.
- Onishi A, Iwamoto M, Akita T, Mikawa S, Takeda K, Awata T, Hanada H, Perry A 2000. Pig cloning by microinjection of fetal fibroblast nuclei. *Science* **289**(5482), pp. 1188–1190.
- Solter D 2000. Mammalian cloning: advances and limitations. *Nature Reviews Genetics*1, pp. 199–207.
- Wilmut I, Schnieke AE, McWhir J, Kind AJ, Campbell KH 1997. Viable offspring derived from fetal and adult mammalian cells. *Nature* **385**(6619), pp. 810–813.

In addition to using a conventional heading sequence it is also helpful to use a design grid, as shown in Figure 11.18. A design grid is simply a series of horizontal and vertical lines on the page that are used for lining sections up; the lines are then removed when the poster is printed. Most drawing software has a design grid feature, or alternatively you can draw the lines manually with a line tool to create the grid and remove them before printing.

In the above examples the methods section is unusually long. However, remember that this is a poster summarizing a current research technique. More commonly the methods section would be much shorter, leaving space for fuller results, discussion, and conclusion sections. In spite of the unusually long methods section, however, the above layouts are fairly conventional, following a symmetrical three-column format. There are many other alternatives though, and you will need to choose the most appropriate one to fit your content. Other examples for both landscape and portrait layouts are shown in Figures 11.19 and 11.20.

FIGURE 11.19 Alternative landscape layout.

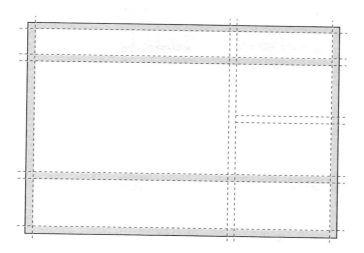

FIGURE 11.20 Alternative portrait layouts.

BOX 11.2 Do the small details really matter?

Posters can be very fiddly and therefore time-consuming to create. However, remember the academic context of poster presentations that we summarized in section 11.1: in a busy conference setting with lots of posters to view, people make very quick (often unfair) judgements about whether or not they will bother viewing a poster. We said that a poster needs to stand out from the crowd in some way if it is going to be viewed. Positively, this will include a snappy title, good visual impact, and information that is readily understandable; negatively, small errors in the design can put people off and lead them to think that if the design isn't accurate the data might not be either. Also, if you are creating a poster at A4 size, then enlarging it to A1 or A0, any small errors at A4 will be enlarged 300% or 400% respectively. Therefore, attention to detail is important. See Figure 11.21 for an example of how small inaccuracies with the layout at the A4 stage lead to much larger inaccuracies at A0.

FIGURE 11.21 How small errors are noticeable when the design is enlarged to poster size.

11.2.6 Choose a colour scheme

The penultimate step is to choose a colour scheme. Try to use only two or three different colours, plus black, which is always best for the smallest text. There are two ways of choosing your colours: you can use colours from your images or a colour wheel.

Colours from images

One way of selecting colours is to use one that is represented in your images. So for the poster we have used as an example we might choose a pink because of the pig photograph, or a brown because of the cow photograph.

Colour wheel

The alternative to choosing colours from images (either because you don't have any appropriately coloured images or you are not confident in choosing from them) is to choose colours from a colour wheel. A colour wheel, as shown in Plate 1, shows the primary, secondary, and tertiary colours and can be used to choose appropriate mixes of colour.

You can choose either:

- analogous colours (colours next to each other on the colour wheel);
- complementary colours (colours opposite each other on the colour wheel);
- or shades (different shades of the same colour).

11.2.7 **Check it very carefully**

Finally, you need to check your poster very carefully. This is particularly important if you are going to get an enlarged, laminated copy printed: you don't want to spend money printing it out only to find the errors afterwards. A good way to check your poster is to print it out A4 (preferably in colour), stick it on the wall, and take a step back to look at it. This will simulate more closely how the real thing will be viewed and help you find any errors more easily. Alternatively, if you have access to a data projector, you can project the image onto a screen to see the effect of the image being blown up to full size (although the image resolution is much less for digital images than printed images, so don't use a projected image as a guide for image quality).

✳ **Chapter summary**

Designing academic posters can be a difficult and time-consuming task. However, by understanding the purpose of posters and the key steps involved in creating them, designing and producing quality academic posters becomes an achievable goal. Designing posters is a particularly useful skill if you continue studies beyond undergraduate level, as they are a common form of presenting research at academic conferences.

Chapter 12

Using feedback

⊃ Introduction

So the essay you worked so hard on has been marked and returned to you. You take a quick look at the mark and that's quite good, so you put the essay down among the pile of paperwork on your desk and it doesn't get looked at again. You feel quite happy but you have also just denied yourself a valuable learning opportunity: you got quite a good mark but why wasn't it really good? What could you have done better? How can you get a really good mark next time?

This is where feedback, and how you use it, is important, and is a vital component in the foundational skill of developing yourself that we identified in the first chapter. It is all too easy to make excuses for not doing anything: 'I'm too busy at the moment, but will look at it when I get the chance'; 'I did the piece of work so long ago, there's no point in going back over it now'; 'We've finished that module, so there won't be anything useful anyway'; 'I never understand what Dr Blogs writes, so there's no point in looking at it'. No doubt you can come up with quite a few more seemingly plausible reasons for not bothering to revisit the piece of work, maybe not even bothering to collect it in the first place. However, whatever form the feedback takes, even if it is really limited and you only have the bare marks, you can still make use of it to improve for next time.

12.1 What is feedback?

Ideally, feedback should serve three key functions:

1. Provide you with a clear indication of how well you are progressing in your work in relation to the standards established by your university.

2. Give guidance as to how you can improve for when you come to prepare the next piece of work.

3. Flag up things you are doing well and should therefore continue doing.

At its most basic level, then, feedback takes the form of the marks you get for the piece of work: these marks give you an overall indication of how well you are doing. In terms of specific guidance, marks on their own are clearly not very helpful; nonetheless, you can still use them to help you improve (see *How to make use of feedback*, below).

One level up from this in terms of usefulness is negative feedback that indicates what you got wrong or did badly. This is probably the most common form of feedback and it may take the form of specific comments such as: 'you did not discuss the role of . . .' or 'this is wrong'. The marker may also include more generic comments such as 'the essay was poorly structured' or 'not enough detail'. These types of comment are very common, in part because they are usually the easiest type of comment for a marker to make. By pointing out what is wrong, the marker is providing limited guidance as to areas for improvement, but it is often not clear exactly what you should do to improve, particularly in the case of the generic comments that may be quite vague. In this context, therefore, the best negative feedback is that which identifies the error and then provides guidance as to how to overcome it. This may take the form of full examples, or perhaps pointing you in the direction of specific resources that can help you:

> 'Make sure that you use the appropriate referencing style for your essays—look at the section on referencing in the course study guide.'
>
> 'You have misunderstood the mechanism of oxygen transport—see pages 32–35 in your physiology course text.'

A less common form of feedback is positive feedback. Here the marker not only provides encouragement: 'this was a good essay' but also identifies specific aspects of the work that were well done:

> 'This was a very good essay and I particularly liked your use of diagrams to support the arguments you were making.'

As with the negative feedback, the value of positive feedback is greatly increased where examples are given so that you are clear about what things you should carry on doing. Inevitably, the quality and level of detail of the feedback you receive will vary. Indeed, recent National Student Surveys have all identified feedback as the aspect of higher education that students find less satisfactory. However, it must be remembered that feedback is a two-way process and you will only benefit from it if you engage with the process. In this way, even if the feedback you receive is not always very detailed or is late in coming, you will still be able to identify ways of improving what you do.

12.2 When do you get feedback?

Many students only think of feedback in terms of written comments on coursework assignments. Indeed, this is the most obvious and one of the most common forms of feedback. However, feedback comes in a variety of forms and, unless you are ready to recognize when you are receiving feedback, you will be missing out on valuable learning opportunities. Let's now consider the range of different feedback types that may be available to you.

12.2.1 Written comments on written assignments

As stated above, written comments on written assignments constitute the most obvious form of feedback and are probably the most *common* form of formal feedback. These types of comments usually take two forms:

1. Generic comments, usually written at the end of the piece of work, or on a separate cover sheet. These comments tend to relate to the overall style and quality of the essay.
2. Specific comments which usually take the form of annotations made on the script itself. These may relate to stylistic issues, such as paragraph structure or referencing, or they may be subject specific, such as the identification of a factual error or the omission of a point in the argument.

12.2.2 Comments on oral presentations

These feedback comments may be given in the form of written comments on a mark sheet, but they are often given just as verbal feedback following the presentation. Such verbal feedback can be very useful because it is the most immediate form of feedback, being given within a few minutes of the delivery of the piece of work. However, verbal feedback is also transient: if you don't write the comments down you are likely to forget them very quickly and then you won't derive any long-term benefit from them.

12.2.3 Comments during tutorials

Tutorials are often a source of very useful discussion about specific topics in the curriculum (see Chapter 7, *Working in tutorials and groups*). However, as with the verbal feedback on oral presentations, the points made by a tutor will be transient and quickly forgotten unless you note them down.

12.2.4 Comments during practical classes

Practical classes offer another very useful way in which students can engage in discussion with tutors or demonstrators. As a consequence, they can be a rich source of feedback, for example in relation to experimental technique, analysis and presentation of results, or in understanding of the topic itself. Again, however, much of this feedback is transient and it is very important that you note the comments before they are forgotten.

12.3 How to make use of feedback

If you are going to benefit from feedback that is provided, then you must take steps to engage with it as an active process. This doesn't just mean looking briefly at the comments. Engagement requires a detailed reflection of how the comments relate to the piece of work you have submitted and consideration of how you should change your approach to future pieces of work.

There is often a perception that if feedback is not provided very quickly, then it is of no value. The usual reasons for taking this view are:

- that it is several weeks since you did a piece of work and you can no longer remember much about it; and

- that the module for which you did the piece of work has now finished and therefore the feedback no longer appears relevant.

Both these reasons, while initially understandable, are false. There are always useful pointers that you can take from feedback, though it may be that sometimes it requires more work on your part to extract the maximum value.

For the most part in this section, we will focus on making use of written feedback since this is the most common form of formal feedback on coursework; however, much of the guidance also holds true for verbal feedback. The difference between the two is that in the latter case, it is up to you to make sure that you make a record of the feedback.

12.3.1 Engaging with written feedback

If you have a cover sheet, as is used by many universities, or if you have been given summary written comments at the end of the essay, then read these carefully before doing anything else. Divide the comments into two groups.

1. Generic comments: ideally these take the form of 'feed forward' comments such as identification of broad areas for improvement. This could include how to improve your overall writing style, to reference better, or to draw graphs and present data better. There may also be positive comments, identifying things that you have done particularly well and should carry on doing in future work.

2. Subject-specific comments: these will normally be related to factual information such as the identification of factual errors or omissions. Again, markers should also flag up good points, so you get a feel for what you are doing well and for what needs improving.

Having categorized the comments, read through the piece of work carefully; after all it may be some time since you wrote it. As you read through, bear in mind the comments that were made and try to be objective in your evaluation. Also make sure that you can identify what aspects of the essay or report the comments relate to. You should also note any annotations on the script. With the generic comments, make sure you understand how the comments apply to your work and how you can improve what you have done. Make a note for future reference so that you can carry them forward to your next assignment: although the next assignment may well be for a different module, generic issues such as referencing style will still be relevant. Likewise, make a note of aspects of your work that were highlighted as being particularly good. With subject-specific comments, check that you understand any factual errors or omissions. Make a note to go alongside your lecture notes to make sure that the point is clear for revision purposes.

Look at your marks, and check them against any marking criteria that you might have been given for the assignment or against any generic criteria that your department might have published. Make sure that you appreciate how your work fits against the criteria and why it falls into the specific mark band that was allocated. Again, if you don't understand, then ask.

For example, if you obtained a mark of 55%, the marking scheme might state that the work has the following characteristics.

> The essay is well organized, displaying understanding of the main issues. There may be a few, minor errors or poorly expressed ideas. There is a significant dependence on lecture notes and/or textbook material.

As you read through your essay, make sure you can identify why the marker placed it in this category: were there some minor errors? Did you mainly use lecture notes or textbooks rather than research literature (see Chapter 3, *Working with different information sources*)? Think about how you could do it better next time.

Feedback may not always be clear! For example, the comments may be illegible or perhaps you don't understand how to improve even though the marker has highlighted a weakness. You might not understand what the marker means by some comments. If you are not sure what to do, then do ask your tutor to explain in more detail how you can improve.

Make sure that you take the feedback forward and use it to improve your next piece of work and your understanding of the subject. When you have taken the points forward, also check the feedback on the next piece of work to ensure that you have indeed improved.

12.3.2 Engaging with oral feedback

As indicated earlier, this may take various forms, for example in the laboratory, a tutorial, or as feedback on a presentation. The danger is that this form of feedback is very transient and easily forgotten, therefore you must be sure to take notes so that you can remember the specific points. The principles, though, are the same as for written feedback: divide the feedback up into generic and specific aspects, and make sure that you understand what the issues are and how to take them forward.

Feedback, then, is one of the keystones of good learning, but you can't treat it as a passive process, otherwise you will not progress and you will make the same mistakes again and again. Engagement is all-important.

✳ Chapter summary

Figure 12.1 shows a flow chart of the processes of using feedback. The most important aspect of this is that you need to think about the feedback, see how it relates to the work that you have had marked, and then to see how you can use it in future to improve your performance.

FIGURE 12.1 Engaging with feedback.

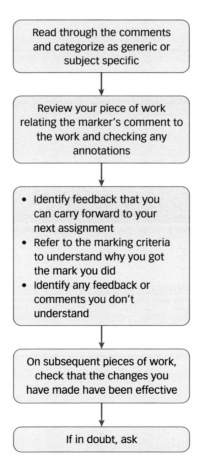

Chapter 13

Getting the most out of revision

➔ Introduction

The results you gain for your degree will depend on how you perform in the various types of assessment that you are required to undertake. We have already addressed essays, practical reports, and presentations; the one we haven't dealt with yet directly (and the one that for most degrees accounts for the majority of the marks) is examinations. In order to do well in exams you have to do well at revision (two words that usually conjure up fairly negative emotions in people). This chapter will show you how you can revise well; Chapter 14 will deal with the exam skills.

If exams account for the majority of the marks that make up your degree result, it is understandable that people feel pressured or stressed by them. One important way of significantly increasing your chances of doing well and of alleviating some of the pressure and stress of exams is to get the revision part right. If you can go into an exam confident that you have prepared yourself as well as you could have done (or, at least, reasonably well), then a significant amount of the pressure and stress can be eased, making it more likely that you can perform well. However, getting the revision part right takes a significant amount of organizing; it is no good leaving it to the last minute.

Firstly though, let us consider what revision actually is. In very general terms revision is to do with going over a subject again, usually in preparation for some kind of exam. But exactly how you go over a subject again is critical to whether or not the revision is effective. You will know if your revision is effective because revision that is effective will make the material you are revising easier to:

* understand;
* remember;
* apply.

This chapter will give you some tips and techniques to help you make this happen.

13.1 **Get yourself organized!**

We have already highlighted the fact that getting revision right takes a significant amount of organizing. However, getting yourself organized is easy to say but is more difficult actually to do, especially when you're beginning to feel the pressure of exams looming and you just want to get on with revising. It is important, however, to see the organizing or planning aspects of revision as part of the process, rather than as something to get out of the way so you can get on with the 'real work'. This is similar to how we identified planning writing as an important part of writing essays in Chapter 5. Time spent *organizing* revision is time well spent (as long as you don't use it as an excuse to procrastinate!).

There are several questions you need to ask yourself in order to get organized.

- What will I be examined on?
- Do I have all the necessary material?
- How much time do I have before the exams?
- How am I going to allocate my time?

These questions will form the structure of this first section.

13.1.1 **Find out what you will be examined on**

Firstly, then, you need to find out as much as possible about what you will be examined on. This includes both the format and scope of the exam (or exams). In terms of format this would include finding out answers to the following questions.

- How long will the exam last?
- How many questions will there be?
- Will you have any choice about which questions you answer?
- What type of questions will you be asked (e.g. multiple choice, short answer, or essay)?

Knowing the answers to these basic questions will help you to know what to expect. The better informed you are the better you can prepare. It also means that you are not surprised when you open the exam paper and so will be more able to concentrate on answering the questions rather than spending time trying to understand issues of format.

In addition to the format of the exam you also need to consider its scope. What are you expected to know in terms of both breadth (the range of information) and depth (the amount of detail)? Imagine you were revising for an exam on comparative animal physiology: if you had revised enough in terms of breadth but had revised insufficient depth, your knowledge of comparative animal physiology would be too *shallow*. For example, you may have revised the comparative physiology of the respiratory, cardio-vascular, and excretory systems, ion balance, and vision, but failed to learn them in sufficient detail. Alternatively, if you had revised enough in terms of depth but had revised insufficient breadth your knowledge of comparative animal physiology would be too *narrow*. For example, you may have revised vision in intricate detail, but if you failed to learn about the other aspects as well (respiratory, cardiovascular, and excretory systems, and ion balance), then you would be very limited in the number of questions you could answer.

This may all seem fairly obvious, but finding out what you will be examined on is an important and often overlooked area of revision. Sometimes it is overlooked due to carelessness, other times due to a deliberate but irrational avoidance mentality. Whatever the reason, it is not difficult to fix. There are a number of ways you can find out information about exam format and scope including:

- module handbooks;
- past papers;
- course tutors.

Each of these will give you important insights into what you can expect.

13.1.2 **Check you have all the necessary material**

The second question to ask in order to get yourself organized is 'Do I have all the necessary material?' When you have found out about what you will be examined on (the breadth and depth) you then need to look at the material you have and decide whether or not you have enough. It is important that you think about more than just your lecture notes at this point. Other sources of important information include:

- notes from tutorials;
- essays you have written for coursework;
- lab reports that you have submitted;
- notes from presentations that you have been required to give;
- notes from additional reading for all of the above;
- feedback from tutors on all of the above.

As you can see, the amount of information you will have accumulated over a term or semester will be considerable. At this stage you are not supplementing your notes, simply auditing them to check where there are any gaps. We said earlier that module handbooks were an important source of information on exam format and scope, and as such they help you to identify gaps in your notes and so help you determine whether or not you have enough material. Simply make a list of what information you have on a summary sheet and compare it to what you are expected to know, as described in the learning outcomes for the particular module you are revising for.

When you identify gaps in your material (we say 'when' rather than 'if' because it is unlikely that you will have everything), make a note of what the gaps are so you can come back to them later. It is also worth identifying whether the gap is a breadth issue (an insufficient range of information) or a depth issue (insufficient detail). The other factor to consider is whether certain gaps actually matter: if the format of the exam allows you some choice as to which questions you answer (e.g., some essay-based exams) you may not need to revise everything. It is usually better to have revised most things in sufficient depth rather than everything at just a surface level. It may also be that there were some sections of the module or course that you just could not get to grips with and still don't understand. There are risks associated with this strategy of selective revision, so you need to be careful. For essay-based exams where you have a choice of questions, it is still very important that you revise more topics than you actually have to answer questions on, so that you do have a choice when you come to read the paper. A reasonable rule of thumb is always to learn at least two more topics than you have to answer questions on, so if, for example, your typical essay paper requires you to answer three questions, then it is best to know five topics really well.

13.1.3 **Note how much time you have before the exams**

The third aspect of getting yourself organized is to be clear about how much time you have to revise. Once you know the dates of your exams you will know how much time is available to you. The best way to impress on yourself how much time you have available is to represent the time visually by using a diary, calendar, or simply a sheet of paper. In addition to the exam dates themselves, you need to include other fixed points, such as any remaining coursework deadlines, the final lecture, and any non-course commitments you have, such as holidays. All of this needs to be represented on a planner, such as the one shown in Figure 13.1.

It is unlikely that all of your exams will be at the same time, so you can stagger the start of the revision for each exam accordingly. Knowing when your exams are will help you to appreciate how much time you have before the exams. The next step is to decide how you are going to allocate the time you have.

13.1.4 **Decide how you are going to allocate your time**

Deciding how to allocate your time is an important element of getting yourself organized. The following guidelines will help.

Decide what you are going to revise and how much

In order to decide how you are going to allocate your time you need to decide what you are going to revise and how much you are going to revise it. You will already have some idea of this, having found out what you will be examined on (section 13.1.1) and whether you have all the necessary material (section 13.1.2), but now you need to make

FIGURE 13.1 Planning your time.

April						
		1	2	3	4	5
6	7	8	9	10	11	12
13	14	15	16 Final lecture	17	18	19
20	21	22 Essay due in	23	24	25 Away for weekend	26
27	28	29	30			

May						
				1	2	3
4	5	6	7	8	9	10
11	12	13	14 Exam 1	15	16 Exam 2	17
18	19	20 Exam 3	21 Exam 4	22	23	24
25	26	27	28	29	30	31

some definite decisions about how you will allocate the revision of your material to the time you have available. There are some important considerations to bear in mind at this point:

- if you are going to know your material in sufficient depth you will need to revise subjects more than once;

- however, that doesn't mean you have to revise absolutely everything (remember, it is usually better to have revised most things in sufficient depth rather than everything at just a surface level);

- you might also decide that you need to revise more for some exams than for others—you don't necessarily have to allocate equal amounts of revision time for each exam.

A useful process is to make an overview of each subject, listing the topics in that subject, and the headings under each topic (you will already have most of this information if you have identified whether you have the necessary material). This creates an index of what needs to be revised so that you can divide the revision into easy-to-manage sections. Once you have this overview, you are ready to put together your revision timetable.

Make a revision timetable

The kind of plan represented in Figure 13.1 gives you a useful overview but it doesn't give you enough detail about how you need to allocate your time. You need to zoom in on this overview to create monthly, weekly, and even daily plans of what you need to do.

Imagine your four exams were as follows:

1. Comparative animal physiology: 14 May

2. Human genetics: 16 May

3. Biochemistry of enzymes: 20 May

4. Animal and plant diversity: 21 May

According to your plan (Figure 13.1) this gives you 4 weeks between your final lecture (16 April) and your first exam (14 May). However, there is also your final essay to submit on 22 April, and assuming you will need the time between the final lecture and the submission date to work on the essay—that leaves just over 3 weeks to revise before your first exam. The remaining three exams then fall within the following seven days. These weeks need planning carefully if you are to cover the necessary material in the time available; this is where the weekly and daily plans come in (Figure 13.2).

Be realistic and build in some slack

It is important to be realistic in your planning. Estimating clearly how much time you have should help you not to panic and so end up trying to cram too much revision into a

FIGURE 13.2 Monthly, weekly, and daily revision plans.

FIGURE 13.3 Daily plan with slack built in.

short space of time. If you are clear how much time you have and what you need to do, you can then allocate your time realistically. A realistically planned revision timetable will have some slack built in so that when your plans need to change (for instance due to a certain topic taking longer than expected) there is space for this to happen. Of course, if the slack time is not needed you can use it for other things, for instance going over a topic again in more detail or perhaps rewarding yourself with an unexpected break. It is difficult to be precise about how much slack time is needed, but if you planned to revise 7 hours in a day, you could leave the seventh hour unallocated to use for whatever was necessary. An example is shown in Figure 13.3.

In addition to planning in slack time you also need to plan in breaks: revising non-stop for seven hours is not an efficient use of time. The longer you go without a break, the more difficult it becomes to concentrate and absorb information. Breaks from revision, therefore, are not a luxury; they are a necessity if you are to be able to revise in a sustainable manner. Therefore, the morning and afternoon slots represented in Figure 13.3 shouldn't be 3 hours without a break, but rather shorter blocks with breaks built in. Keep your breaks short and free from distractions; just getting up to walk round the room or make a drink is enough to renew your concentration. You also need to decide what times of day you work best; some people find working in the mornings easier, others find evenings more productive. Whatever your preference is, try to use it to your advantage, but don't use it as an excuse not to get up in the morning!

Do other stuff

Obviously it is important to be focused on your revision, but it is important to have a life outside of revision too. The longer your revision lasts the more important it is to plan in time to do other things. Planning in time for other things isn't simply an end in itself but can be used to help you in your revision in a number of ways, for instance:

- provide you with a break, so you can come back to your revision more mentally refreshed;
- re-energize you, so you can come back to your revision more physically refreshed;
- act as a reward for having achieved a certain target in your revision.

Precisely what you do to achieve these benefits will depend on what you find helpful. People often find exercise helpful, or perhaps spending time socializing. You will need to be disciplined and not get carried away, but a short period of activity, perhaps each day, that you find enjoyable and would achieve some of the benefits listed above will help keep your revision on track and sustain your motivation.

Use variety

In addition to spending short periods of time doing other things, another helpful strategy is simply to use a variety of approaches to your revision. You will know that this is helpful if you have ever tried to spend a revision session just reading, for example. Spending a large amount of time only using one approach can quickly become tedious and de-motivating. But this isn't necessarily a sign that you need a break, it might just be a sign that you need to change your method. Try testing yourself with some questions or flash cards, or drawing a diagram, or writing a summary of what you have just read. We will deal with these, and other, suggestions in more detail shortly when we consider active revision techniques, but the important point to note here is the principle that using a variety of approaches to how you revise can be helpful. A change is often as good as a break (and better than a break if you are using breaks to avoid getting down to revision!).

Go public

Lastly in this section on allocating your time, consider making your revision timetable public; you could stick it up on a kitchen cupboard or on your bedroom door. There are a number of reasons this can be helpful:

- it reminds you of what your plans are;
- it informs others who share your accommodation what your plans are;
- it makes clear to others when you are available and when are not;
- it makes you accountable to others ('I thought you were supposed to be revising?' or 'Shouldn't you be taking a break now?').

Whether or not you want to make your plan public will depend on who you live with, but it is definitely worth considering.

13.1.5 Check that your environment is suitable

Finally, in this section on getting yourself organized, you need to think about the environment in which you will revise. You might be the sort of person who always prefers to work in the same place every time, or perhaps you like variety. Some people find the campus library a helpful place to revise, whereas others prefer to study in their own room. Whatever your preferences try to make decisions based on what will help your revision to be most effective. A suitable studying environment is one which is free from distractions. For example, it should be:

- well lit;
- quiet;
- not too hot or too cold;
- reasonably tidy and free from irrelevant clutter.

Sort these things out quickly before your revision session begins, but don't use them as an excuse to avoid starting!

13.2 Use active revision techniques

We said at the beginning of this chapter that you will know when your revision is being effective because it will make your material easier to understand, remember, and apply; this is what active forms of revision help you to do. **Active** revision is any form of revision that makes you interact with the material in an involved and thoughtful way; this includes condensing your notes, drawing summary diagrams, and testing yourself with questions or flash cards. **Passive** forms of revision, however, such as reading or copying, lack the focus provided by active forms and are more difficult to sustain for longer periods.

So, now you're organized, what are you actually going to do when you sit down for your first revision session? Assuming that you have planned your revision timetable so you have a reasonably clear idea of the topics you will be covering on any given day and how long your revision session for the day concerned will last, you are now in a position to begin to work on the detail. This will involve:

- filling in any gaps in your material;
- condensing your notes carefully; and
- reviewing your notes regularly.

13.2.1 Fill in any gaps

The starting point then is to fill in any gaps in your material you have identified (as described in section 13.1.2 above) and have decided need addressing. Your material could be supplemented from a number of sources: if you have simply missed a lecture, then borrowing a friend's lecture notes may be adequate, but it is more likely that you will need to do some additional reading. This needs to be done in a certain sequence: make sure you understand the basics first before you progress onto more advanced issues. In practice this means you need to start with lecture notes and handouts, followed by key chapters from core texts on your reading list. If you still need more detail then, once you are confident you understand the basics, you could move onto more specialist publications. This is where it is important that you have understood what you will be examined on (see section 13.1.1) so you are aware how much additional reading is necessary. There will always be more reading that you *could* do; the question you need to answer is whether there is more reading that you *must* do? At this stage, you need to be strategic in your approach: you don't have a lot of time to be undertaking detailed additional reading, therefore you need to identify exactly what is necessary and do that, but avoid doing more.

13.2.2 Condense your notes carefully

Once you have filled in any gaps in your notes you will probably have a lot of material. Therefore, a vital stage in any revision strategy is to condense your notes into a format that is more manageable. There are two simple reasons for this:

- the process of condensing your notes helps you to learn the material;
- the end product of condensing your notes provides you with a summary of your material which you can easily review (also enabling you to learn your material).

We will address the review part shortly (section 13.2.3 below), but first we will deal with the condensing element. Condensing your notes is an example of an active form of revision: it forces you to make decisions about which parts of your material are the most important and will help you to check how much you understand. This process consolidates your knowledge and helps you to make connections between different aspects of your learning and identify underlying principles. But how do you actually condense your notes? Condensing notes involves the following stages (and is illustrated in Figure 13.4).

- Take your original notes on a topic (from various sources) and make a condensed version, perhaps on several sheets of paper.
- Then take this condensed version of your notes and try to condense the information still further, perhaps to a single sheet of paper.
- Finally, write an overall summary of the particular topic you are revising, perhaps on an index card.

The principle of condensing is simple enough, but how do you decide what information is sufficiently important for you to record and what you can leave out? Clearly you don't want to simply copy everything out (this would be a passive form of revision which would quickly become ineffective) but, equally, you don't want to be so sketchy with your notes that when you come back to them at a later stage there is insufficient detail for you to work from. This is why this process involves a number of stages: you can't skip from your lecture notes straight to an overall summary without going through the intervening stages. The reason for this is that, in producing condensed notes and then summaries of these condensed notes, you are not only producing a more condensed version of your material on paper but also structuring and connecting the information in your mind and making decisions about what is sufficiently important to record and

FIGURE 13.4 Condensing notes.

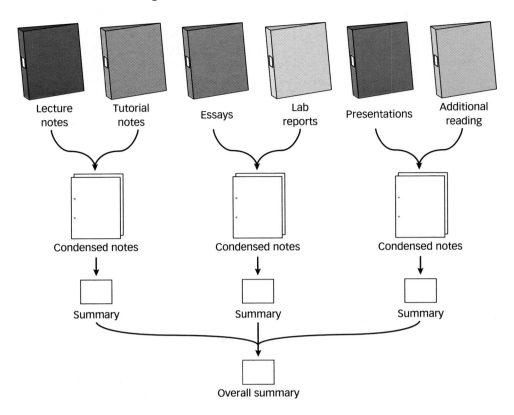

what you can leave out. At each stage you are attempting to distil out the essence of the particular topic you are studying in order to understand it better. What you will be left with, therefore, is not everything you need to know on the topic but sufficient information to prompt you to recall the source information from which it was distilled.

Exactly how much information you record depends on how much you already know and so will vary from individual to individual. What you are aiming for, however, is to pick out key facts and create structures to help you to understand whole concepts. The following example illustrates how this might work.*

Imagine you had the following notes from a lecture on hypoxia.

> Tissues consume oxygen and produce CO_2 at variable rates. The circulatory and respiratory systems must work together to supply and remove gases at an appropriate rate for each tissue of the body. Each step in the chain of supply can be affected by disease, but the ultimate consequence of any such condition is that some or all parts of the body receive less oxygen than they need—hypoxia, sometimes also associated with the inappropriate accumulation of carbon dioxide.

*Thanks to Professor Stewart Petersen from the School of Medicine, University of Leicester, for these examples.

The supply of oxygen to each individual tissue depends first on appropriate oxygen content in arterial blood, and second on adequate local perfusion.

Appropriate oxygen content in arterial blood depends on effective diffusion of oxygen from alveoli into alveoli capillaries, and appropriate partial pressures of oxygen in the alveolar gas.

Appropriate pO_2 in alveolar gas depends upon adequate ventilation of the alveoli, which itself depends upon the ease of air flow through the airways, the capacity of breathing movements to generate appropriate pressure differences and appropriate composition of inspired gas.

Any disease process affecting oxygen supply to tissues must affect one or more of these steps, which therefore provide a convenient classification for pathophysiology, and the ultimate outcomes, local or generalized, are poor supply of oxygen—hypoxia.

A useful way to start to summarize such notes is to annotate them as you read them, as shown in Figure 13.5.

FIGURE 13.5 Annotated notes.

Hypoxia
Tissues consume oxygen and produce CO_2 at variable rates. The circulatory and respiratory systems must work together to supply and remove gases at an appropriate rate for each tissue of the body. Each step in the chain of supply can be affected by disease, but the ultimate consequence of any such condition is that some or all parts of the body receive less oxygen than they need – **hypoxia**, sometimes also associated with the inappropriate accumulation of carbon dioxide.

The supply of oxygen to each individual tissue depends first on appropriate oxygen content in arterial blood, and second on adequate local perfusion. Appropriate oxygen content in arterial blood depends on effective diffusion of oxygen from alveoli into alveoli capillaries, and appropriate partial pressures of oxygen in the alveolar gas.

Appropriate PO_2 in alveolar gas depends upon adequate ventilation of the alveoli, which itself depends upon the ease of air flow through the airways, the capacity of breathing movements to generate appropriate pressure differences and appropriate composition of inspired gas.

Any disease process affecting oxygen supply to tissues must affect one or more of these steps, which therefore provide a convenient classification for pathophysiology, and the ultimate outcomes, local or generalized, are poor supply of oxygen – hypoxia.

You could also choose to summarize the content in your own words, as shown in Figure 13.6.

FIGURE 13.6 Summary notes.

- tissues consume O_2 and CO_2 @ variable rates (circ. + resp. systems together supply + remove gases @ apt. rate)
- each step of supply can be affected by disease → less O_2 than needed (hypoxia)
- supply of O_2 to tissues depends on
  1. apt. O_2 content of arterial blood
  2. adequate local perfusion
- 1 depends on effective diffusion of O_2 from alveoli → alveoli capillaries & apt. PO_2 of O_2
- which depends on adequate ventilation of alveoli
- which depends on ease of air flow thro' airways, breathing movements + comp. of insp. gas

NB. diseases affecting O_2 supply to tissues must affect at least 1 step

To condense the information further you could create a summary diagram, as shown in Figure 13.7. This summary diagram has the added advantage of including additional information too.

FIGURE 13.7 Summary diagram.

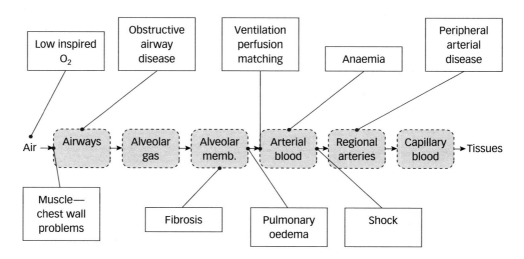

In addition to considering the content of the information you are condensing, you also need to consider its format. The example used in Figure 13.7 is a deliberately visual example. This is because many people find it easier to remember something that is visual rather than something that is essentially text-based. In Chapter 2, *Making the most of lectures*, we suggested a number of points to improve your note-making; many of these are also relevant for note-making for revision. Remember to try to use:

- headings and sub-headings;
- colour;
- space;
- figures, including labelled diagrams and flow charts.

13.2.3 **Review your notes regularly**

We observed above that condensing your notes provides you with a summary that you can easily review. If you have ever tried to review uncondensed notes you will know that it is a very difficult exercise to sustain because of the sheer volume of information you are trying to read. Even if you manage to review uncondensed notes for a particular topic, the more topics you have the more unsustainable it becomes. If you have produced summaries, however, not only do you have your notes in a handy format for you to review, as identified above, but the process of condensing your material will also have helped you to learn it. Clearly you won't remember everything (that is why the review stage is important) but you will remember some and you will also have something that is reviewable.

The importance of reviewing your notes regularly can be highlighted by considering what happens when you don't review your notes regularly. What inevitably happens is that the longer you go without reviewing your material the more you forget. Without regular reviews of what has been revised you won't retain the information you have taken the trouble to learn.

The relationship between reviewing and recalling information is straightforward enough, but how do you actually review your material? There are many different methods of reviewing information but, whichever method you choose, it needs to test your recall of a topic by:

- checking your ability to understand central concepts; or
- remembering key facts; or
- linking the information to other topic areas;
- or perhaps all three!

As we have already noted, you won't remember absolutely everything, and you can always go back to your original notes to remind yourself of particular details; the important thing is to do the reviewing. Methods of review include:

- testing your recall of key facts with index cards (or flash cards), putting to one side the cards you can remember and repeating the cards that you can't;

- reproducing from memory headings and keywords for a learnt topic area and then checking them against your notes;
- recording yourself explaining concepts or theories to an imagined audience, then playing back your explanation, checking it against your notes;
- making a visual summary (like the flow chart in Figure 13.7) to summarize information you have remembered;
- reproducing important diagrams that summarize large amounts of information or whole concepts, for example metabolic pathways;
- enlisting the help of a friend to test your recall using any of the above methods;
- putting up summaries of key information around your home to give you the opportunity for on-the-spot tests;
- testing yourself with question banks and past papers (more on this shortly).

Reviewing your material need not take a lot of time. A good time to review material is at the beginning of a revision session but it can also be fitted into any spare time during the day. Frequently reviewing your material will help to build up your confidence and reassure you that you are making progress.

13.3 See the big picture

We have noted above that condensing your notes supports the consolidation of your knowledge and helps you to make connections between different aspects of your learning and to identify underlying principles. However, when paying attention to detail, it is easy to lose sight of the context and principles. It is important, therefore, that you make a special effort to appreciate the big picture. Try not to get so focused on the details that you fail to see either where the information fits in to the larger context or perhaps a simple principle that underpins it, both of which would make it easier to understand and remember.

The exact parameters of the big picture will depend on how much breadth and depth you are required to know, and this will change depending on the module choices you make throughout your course and the level of study you are at. However, context and principles are important to appreciate and identify because they help you to develop a means of structuring, storing, and retrieving information. They also help because exam questions rarely ask the precise question you had been hoping for; rather they require you to apply your knowledge to different situations.

Imagine walking into a room with papers scattered across the floor and then being asked to find a particular piece of information—it would be very difficult! If you did manage eventually to find the correct piece of paper it would be difficult to then explain how the information on the piece of paper relates to all the other information on the pieces of paper still on the floor. However, imagine the same papers in the same room but this time organized appropriately into a filing cabinet. Suddenly it becomes much easier to find the required information and relate it to other information. Seeing the big picture is a bit like storing papers appropriately in a filing cabinet (or information in a database): it enables you to appreciate how information fits into a wider context and to identify the principles that underpin it. Without these elements the information you are trying to learn will remain abstract and unrelated, which makes it much more difficult to understand, remember, and apply. It is important therefore that you spend some time standing back from the detail to think about structure. Appreciating context and identifying principles are two specific ways of doing this, but anything which helps you to reflect on and question the information you are revising, helping you to understand themes and relationships rather than detail, will help you to see the big picture.

13.4 Practise outputting the information you have learnt

We have noted the importance of understanding, remembering, and applying the information you learn. However, the steps we have focused on so far are largely concerned with the first two elements: understanding and remembering. If you are going to perform well in the exam room (more on this in the next chapter) then you need to be able not only to understand and remember information but, importantly, you need also to be able to apply it. This is because exam questions, whatever the format, are specific and you are therefore required to give specific answers. When you open an exam paper you are very unlikely to find a question which reads 'Please tell us everything you know about comparative animal physiology'. A much more likely question would be 'Compare and contrast the retinal organization of a named mammal and reptile'. Therefore, the information you have taken time to understand and remember needs to be applied to the specific question. It would be very unfortunate if you had taken the time and effort to understand and remember information but failed at this last crucial step. The best way to improve your ability to answer questions specifically is to practise. The two main ways of practising are:

- creating question banks; and
- using past papers.

13.4.1 Create question banks

A question bank is simply a collection of questions that you think you might be examined on. The questions shouldn't be full essay questions, but short, one-line questions that test very specific knowledge. Question banks are useful for a number of reasons:

- past papers are not always available to you;
- the format of the exam might have changed;
- thinking up questions helps you to reflect on how much you know;
- answering the questions helps you to apply your knowledge.

Question banks can be created from any questions that you think you might need to know the answer to. A good place to start is the learning outcomes for the modules: use these as a prompt to help you think of questions. Keep adding questions to your question bank throughout your course; questions prompted from lectures, or tutorials, or reading. You can then use these questions to check your knowledge of the material and so monitor your progress, or as a means of review, or to supplement your use of past papers. It is also beneficial to team up with some fellow students to share the task of creating a question bank; this will save you time as individuals and give you a greater variety of questions than you might have created on your own.

13.4.2 Use past papers

Past papers are an extremely helpful resource when you come to practise outputting the information you have learnt. If you have already found out what you will be examined on (section 13.1.1) you will know what the format and scope of the exam will be. Assuming there are no major changes to an exam's format or scope for that particular year, you can assume that the past papers will be representative and so be a good test of your learning so far. There are a number of ways you can use past papers, including:

- practise analysing questions and setting up plans to answers in order to test your recall and ability to adapt material to the set question;
- practise writing full answers in the time allowed;
- review your answers by checking them against your notes and highlighting any missed or inaccurate information;
- ask one of your lecturers to read your essay and comment on it;
- analyse your answer to see how it could have been improved (Did it need more information? Did it have a logical structure? Was your expression clear? Did you stick to the question?).

Your precise use of past papers will depend, in part, on the format of the exam: multiple choice, short answer, or essay. We will deal with this in more detail in the next chapter.

✳ Chapter summary

We said at the beginning of this chapter that effective revision makes the material you are revising easier to understand, remember, and apply. Getting yourself organized in your revision, with a clear timeline leading up to the exam, is a crucial first step in revising effectively and can reduce the amount of stress that exams cause. Making your revision an active, rather than passive, process will make your revision both more interesting and more effective. Making sure that you don't lose sight of the big picture in terms of how detail fits into an overall structure is also important, as is practising outputting the information you have been learning to ensure you can apply it to specific exam questions.

Chapter 14

Getting the most out of exams

⮲ Introduction

We ended the last chapter on *Getting the most out of revision* by addressing the use of question banks and past papers as a means of practising how to output the information you have learnt. This final chapter on *Getting the most out of exams* will address the skills you need for the exams themselves. However, exam skills aren't just needed during exams; there are some important exam skills you need to develop for the day or so leading up to exams too. With this in mind, this chapter will begin by addressing what you need to do immediately before your exams, and then progress to give guidance about the exams themselves.

Remember: unlike most other topics covered earlier, you won't have this book to hand for guidance when you're actually experiencing the topic of the chapter—that is, when you are sitting in the exam room. So it is worth spending time reading carefully to make sure you remember all the key points, so that you have them in mind and ready to put into action, when exam time comes.

14.1 Just before an exam

If you have got yourself organized and managed to do a reasonable job of your revision, as described in the previous chapter, then there will be less of a temptation to do any last minute cramming. You need to be realistic at this point though: there will always be more you could do, but at some point you need to stop. It is very unusual, even for students who have managed to get themselves very organized, to be able to walk into an exam room feeling completely prepared (after all, exams are—by their very nature—an unknown quantity; the only way to be 100% prepared would be to know the questions before the exam!). It is therefore common to want to try and cram in a lot of revision just before an exam (either because your revision hasn't gone as well as you had hoped or because you are anxious), but if you do find yourself in this situation then be careful! There is work that is helpful to do at the last minute but, equally, some which can be unhelpful.

14.1.1 **The wrong sort of last minute work**

Last minute cramming is the wrong sort of last minute work. There are a number of reasons for this.

Last minute cramming is ineffective

If there is material that you still haven't covered the night before the exam you are unlikely to gain much benefit by covering it at this late stage. Your time would be much better spent going over things that you are already familiar with (see section 14.1.2 below).

Last minute cramming makes you forget other stuff

Last minute cramming is often counterproductive because trying to cram in a lot of information (particularly new information) at the last minute can cloud your memory of what you already know and so has the effect of pushing out other information that you may have otherwise been able to retain.

Last minute cramming makes you tired

Exams are hard work and require a lot of energy and concentration. Therefore, making yourself tired by staying up late the night before (or worse, through the night) is not good preparation. Your time would be much better spent trying to get some rest so that you are alert for the exam and so can make a good attempt at answering the questions, rather than being so tired you can't think clearly.

Last minute cramming makes you worry

Last minute cramming tends to make you focus on what you don't know rather than what you do know. Thinking about all the things you don't know the night before an exam is a bad idea! It makes you worry, it gets you stressed, it makes it difficult for you to get to sleep, which makes your more worried and more stressed . . . and so it goes on.

14.1.2 **The right sort of last minute work**

While last minute cramming is the wrong sort of last minute work, there is work you can do just before an exam which is helpful.

Plan your use of time

It is important to plan how you will use the time you have in the exam *before* you get to the exam room. This doesn't have to be a last minute exercise, but, if you haven't done it as part of your revision strategy, now is the time to do it. You need several pieces of information to plan your use of time effectively:

- how long the exam lasts;
- how many questions the paper contains;
- how many of the questions you are required to answer;
- whether all the questions are worth the same number of marks.

Once you know this information you can decide how much time you will spend on each question, remembering to allocate time for choosing questions (if you have a choice), analysing questions, planning answers, and reviewing answers, as well as (of course) time to write your answers. For example, if you had a 1-hour short answer paper containing nine questions, each of which was worth 10 marks, you might work out how to allocate your use of time as follows:

- 1 hour = 60 minutes
- 5 minutes per question (including analysing and planning time) = 45 minutes
- which leaves 15 minutes checking time
- 15 minutes checking for 9 questions = just over 1.5 minutes checking time per question

If you had a 3-hour essay paper containing 10 questions, all worth 25 marks, of which you had to answer three, you might allocate your use of time as shown in Figure 14.1.

Planning your use of time before the exam will give you one less thing to think about in the exam itself. It will also ensure that you at least attempt all the questions you are required to attempt, rather than simply focusing on the ones you are confident on. In an essay paper where you are required to write three essays, for example, you will usually

FIGURE 14.1 Planning your use of time.

Time (minutes)	Task
5	Read questions
5	Choose questions
50	Question 1
	– analysing (5)
	– planning (5)
	– writing (30)
	– reviewing (10)
50	Question 2
	– analysing (5)
	– planning (5)
	– writing (30)
	– reviewing (10)
50	Question 3
	– analysing (5)
	– planning (5)
	– writing (30)
	– reviewing (10)
20	Checking

get more marks by having a go at all three essays rather than spending time trying to make two of them really good ones. Even if you are not confident about the third essay and think you can only write a fairly poor answer, the chances are your time would still be better spent attempting an answer rather than refining an existing one.

The night before the exam

We have already noted that the night before an exam is not a good time for learning new information. It is, however, a good time for practising your exam technique or checking your knowledge of what you have already learnt. So use your question banks and past papers to go over the information you have already covered, making sure that if you come across something that you don't know you look it up so you do know it. Going over pre-revised material in this way the night before an exam will both boost your confidence regarding how much you do know (you might be surprised!) and act as a reminder of the things that you had learnt but couldn't recall in response to a question.

The morning of the exam

The morning of the exam is also not a good time for learning new information. As we have seen, cramming makes you forget other stuff; in particular it can cloud your ability to remember overall concepts. Instead, focus on reviewing main points to give you an overview of what you have learnt. If your exam is in the morning, it would be a good idea not to do any sort of revision on the morning of the exam at all.

14.1.3 Other last minute preparations

Good exam technique is about more than simply learning information and then out-putting that information in response to specific questions. Your preparation could have gone brilliantly, but if you are so tired you can't think clearly, or you turn up at the wrong location, your preparations will have been wasted. So here are a few more last minute preparations to consider.

Get your stuff together

Getting everything you will need for the exam organized the night before will give you one less thing to think about on the day itself. You will need a pen you can write with comfortably for a long time, and at least one spare. Depending on the exam you may also need other equipment, such as a calculator. Also think about what you don't need, which is anything that might cause your invigilator to suspect you of cheating if they found it on you. This includes mobile phones, MP3 players, scraps of paper with notes on, and scribbles on the back of your hand.

Double check when and where you need to be

Another thing you don't want to be worrying about on the morning of the exam is when and where the exam will take place. Make sure you have the latest information on the date, time, and location of your exam (sometimes these change due to timetable clashes, which is why the *latest* information is important). Then double check this with other people who are taking the same exam, just to make sure. If your exam is taking place in a location you are not familiar with, it is worth either going to the venue beforehand or arranging to go with someone on the day who is sure they know where it is. Also, set your alarm to make sure you wake up in time, and then set another alarm, just in case. You may also want to let the people you live with know when you need to be up, so they can help make sure you're up too. Having more than one wake-up call reduces the chances of you worrying about not waking up in time and so makes it more likely that you will get to sleep. Plan to arrive at the exam room in plenty of time, as racing in at the last minute (or late) flustered and out of breath is not a good way to start.

Get some sleep

We have already identified that last minute cramming isn't helpful, but a good night's rest the night before the exam is. However, this is easier said than done, so you need to have a strategy. Firstly, make sure that you don't drink too much caffeine too late, as clearly this will keep you awake. Secondly, it is very unlikely that if you stop working at 11 p.m. and then roll straight into bed you will fall asleep quickly, and the longer it takes you to fall asleep the more worried you will become about not being able to sleep and so the more awake you will become.

To help you relax and sleep well you need to plan something relaxing to do in between stopping work and going to bed. It could be chatting to your housemates, going for a walk round the block, or watching a bit of television (as long as you don't start a whole film!). Anything that will help stop your mind racing and calm you down a little will make it more likely that you will be able to sleep.

Manage your stress levels

We have deliberately given this section the title 'Manage your stress levels', rather than 'Get rid of stress' because, firstly, it's unlikely that you will be able to get rid of stress altogether, and, secondly, some stress is helpful: it is your body's way of preparing itself to cope, making you more alert and attentive. All that we have covered so far in this chapter, and the previous chapter on revision skills, will help you to manage your stress levels, because one of the most stressful things, as far as exams are concerned, is lack of preparation. So getting organized will really help: the most stressful exams are always the ones you haven't got organized for. If you still feel anxious though, it can also be useful to talk to someone about your worries concerning exams. This could be a friend, your tutor, or someone in your university's student support services. Additionally, on the day of the exam itself, there are a number of things that will help you manage your stress levels:

- try to avoid talking to other students if you think if will increase your anxiety levels;
- get plenty of fresh air—exam rooms are often stuffy places, so get plenty of fresh air beforehand (this will also help you avoid talking to other students), and breathe deeply to try to calm your nerves;

- make sure you use the toilet before the exam—you really don't want to be needing to go half-way through the exam: it will be a major distraction, and will make you feel unsettled (and, therefore, stressed).

Exams are physically tiring. So there are also a couple of things not directly related to the management of stress that you should keep in mind to make yourself as prepared as possible from a physical point of view.

- Try to eat breakfast—not everyone finds this easy, but exams are hard work and you need energy. At the very least, have something simple but high-energy, like a banana.
- Drink water—to keep you hydrated and help you think clearly (and take a bottle of water in with you).

14.2 Arriving in the exam room

When you arrive in the exam room there are several things you can do to get yourself off to a good start.

14.2.1 Get settled

Exams have a lot of build-up and there is often a lot depending on them. So it is inevitable that you will be feeling at least a little anxious. Just by sitting down, taking a few deep breaths, getting out your pens, your clock or watch, and your bottle of water, and generally getting yourself comfortable in your chair will help you settle down.

14.2.2 Read the instructions very carefully

Once you have been told by an invigilator that you can turn over the paper, turn it over. The first thing to check is that it is the paper you were expecting (if it isn't, put your hand up—often exam rooms contain several different exams at once; you could just be at the wrong desk). Once you are sure you have the right paper begin to read the instructions carefully. The instructions should be familiar to you from your preparation, but check in particular the number of questions, how many questions you are required to answer, and how many marks each question is worth.

14.2.3 Remind yourself of your time plan

You should have planned your use of time beforehand (see section 14.1.2 above), so this should be simply a case of reminding yourself what you had decided to do (but if you haven't—do it now). You may find it helpful to make a note of your time allocation on the question paper so you don't forget or become confused about what you had planned to do (which is easily done when you are under pressure).

14.3 **Answering the questions**

The areas we have covered so far in the sections on *Just before an exam* and *Arriving in the exam room* apply equally to all types of written exam, whether multiple choice papers, short answer papers, or essay papers. However, when it comes actually to answering the questions there are some exam techniques that are particular to different types of written exam. We will deal first with some general advice applicable to all question types and then address how these apply specifically to different types of exam paper.

14.3.1 **General advice for all question types**

Regardless of the type of exam paper you are faced with it's a good idea to have a clear sequence of steps in mind to help you answer the questions to the best of your ability. We suggest the following:

- analyse the questions;
- plan your answers;
- write your answers;
- review your answers.

Following these four steps will help ensure that you have a measured approach to what can otherwise be quite a panicky situation.

14.3.2 **Multiple choice papers**

Of the question types we have identified, multiple choice papers require the least amount of analysis and planning. However, that's not to say you can dispense with analysis and planning altogether—you just need to approach it in a slightly different, and briefer, sort of a way.

Analyse the questions

Usually with multiple choice papers you will not have a choice of questions you can answer. If this is the case then there is no need to read through the entire paper before you begin. Instead, simply read the instructions carefully (as described in section 14.2.2 above) and make sure that the structure of the paper is what you expected. Once you have confirmed this you can begin to work on the individual questions. With multiple choice papers simply start with the first question and then work through the questions in sequence. When you come across a question you don't know the answer to, don't dwell on it too long, just move on to the next one and come back to the ones you are not sure about at the end.

Plan your answers

You neither have, nor need, much time to plan your answers for multiple choice papers. However, a few moments spent planning will help make sure you don't rush headlong into a wrong answer. A good approach is as follows:

- read the question carefully;
- don't look at the possible alternatives yet—try to answer the question independently of the options given;
- look down the list of options to see if your answer is there;
- read all the other alternatives just to make sure.

Write your answers

For multiple choice questions this is simply a case of ringing a response or ticking a box, but do make sure that you have read the question carefully. Sometimes multiple choice papers include multiple response questions as well as multiple choice questions, that is to say questions that require more than one answer. For example:

Which of the following statements about the neuron is true?

(a) The resting potential is dependent on the membrane permeablity
     to Na^+ ions   ☐

(b) The resting potential lies close to the equilibrium potential for K^+ ions ☐

(c) The membrane is only permeable to Cl^- ions   ☐

(d) The membrane is impermeable to K^+ ions   ☐

In this question, as indicated by the stem: 'Which . . . *is* true?', there is only one correct answer (b).

In some questions, there may be more than one true statement. These are indicated in two ways. For example:

Which of the following statements about the lungs are true (tick all that apply)? . . .

Or some take a more complicated form as follows:

Which of the following statements about the ventilation of the lungs is true?

(a) The expansion of the lungs is mainly as a result of contraction of the
     diaphragm.

(b) Expansion of the lungs has to overcome the resistance due to surfactant.

(c) Expiration at rest is the result of contraction of the intercostal muscles.

(d) The tidal volume is typically 500 ml.

    (a), (b) and (c) are true   ☐

    (a), (b) and (d) are true   ☐

    (b) and (c) are true   ☐

    none of the above   ☐

This second type of question is quite complicated and you need to work through the statements carefully, marking each one that you think is true and then seeing how your list matches with the options given.

Also, make sure you have understood the sense of the question. For example, check whether the question is asking you to indicate which is true or which is *not* true.

Your exam questions may also be set out as Extended Matching Sets (EMS). Here you will be given a series of questions, each with statements that need completing; alongside them will be a table of words that could fit in the questions and each word can be used once, more than once, or not at all. A simple version of an EMS question is given below:

1. During quiet respiration, the main contributor to inspiration is contraction of the .... During forced expiration the passive relaxation is supplemented by contraction of the .... and the .... muscles.

2. In actively metabolizing tissues, the pH is .... and the oxygen dissociation curve shifts to the right, leading to .... affinity. This is termed the ....

3. During heavy exercise, the arterial pCO_2 may be .... as a result of the ventilatory response to the production of ....

4. Taking several deep breaths prior to diving extends underwater endurance because the..................... of the blood is initially.....................

External intercostal

Pectoral

Bicarbonate

Increased

pO_2

Bohr shift

Abdominal

Unaltered

Here you have been given a number of incomplete statements and a table of words that could fit those statements. You have to complete the statement with the most appropriate word, bearing in mind that each word may be used once, more than once, or not at all. Again, the best strategy is to read through the statement and think what should go in the space first, then look at the list and see if any of the words match. That way you are less likely to be confused by the distracters.

Review your answers

First, check that you have answered all the questions, and then review your answers, starting with the ones you were least sure about. If you amend an answer make your change very clear; there will be advice on how to do this in the instructions at the beginning of the paper.

14.3.3 **Short answer papers**

Short answer papers can also be tackled using the same four steps. Again, one of the keys to doing well on these papers is being strict with your timing—don't spend a lot of time thinking about each question. If you know the answer, answer it quickly, if you don't know the answer, move on and come back when at the end, when you may have time to think about it more carefully.

Analyse the questions

As with multiple choice papers, short answer papers often don't give you a choice of which questions you can answer. Again, if this is the case you don't have to read through the entire paper before you begin. Start with the first question and read it carefully to make sure you understand it, perhaps underlining key words to help you focus on the exact meaning. If after a quick reading you are not sure how to answer it, move on to the next question and come back to the ones you are less sure of at the end.

Short answer questions usually require specific factual information for the answer, and not detailed description. Usually the number of marks to be awarded is given on the exam paper, so that can be a guide to the amount of information you need to give. Typical short answer questions might take the following forms:

1. Draw a labelled diagram of the nerve action potential (4 marks)
2. Label structures A–D on the diagram of the ear (4 marks)
3. List the stages involved in the replication of DNA (5 marks)
4. Describe the process of synaptic transmission (5 marks)

Plan your answers

With short answer questions, in briefly planning your answers you are trying to make sure that you select information relevant to the question and put it into a logical and coherent order. Thus, in each of the questions above, the marker is looking for very specific factual information. To answer the question, you will need to note down a few keywords to stimulate your memory and organize your thoughts.

Write your answers

Use your brief plan to prompt you of your points and structure your answer. You only have a short time to write each question, so the important point is to make sure that all the key facts are included. Thus, the diagram of the nerve action potential does not need to be a great work of art, but you must remember to include key details such as scales to mark the voltage and the time. When writing your answers, brevity and factual content are the keys to a good answer.

Review your answers

Check that you have answered every question that you needed to answer (probably all of them), then review your answers. As with multiple choice papers, start with the questions you are least sure about because these are probably the ones you will be able to make the most improvement to.

14.3.4 **Essay papers**

Often with essay papers you will have a choice as to which questions you can answer. If this is the case then make sure you read through all the questions carefully before you choose which one (or, more likely, ones) you are going to answer. As you read the questions, mark the ones that you think you will be able to answer the best, then go back and read them again to make a final choice. As with multiple choice and short answer questions, start with the questions you are most confident with to get you off to a good start and stimulate your thinking.

Analyse the questions

Analysing questions is particularly important for essay questions. With multiple choice and short answer papers your answers will usually just be facts or isolated pieces of information. With essay questions, however, there is more work to do to at the analysis stage in order to be able to organize your answers logically and coherently. As we said in Chapter 13, *Getting the most out of revision*, essay questions will usually be very specific. So, instead of asking 'Please tell us everything you know about comparative animal physiology', they are much more likely to ask you to, for instance, 'Compare and contrast the retinal organization of a named mammal and reptile'. To apply your knowledge to this particular question requires you to analyse the question briefly, using the same technique that we identified in section 5.3.2 *Analyse the question or title*. The key information to identify in the question or title is:

- the subject of the question;
- the instruction;
- the key aspect;
- other significant words.

Plan your answers

Planning answers is particularly important for essay questions because it helps you to:

- select information that is relevant to the question;
- put this information into an order that is logical and coherent;
- write your ideas down at an early stage to help you remember your key points;
- monitor how much information you are covering in the time allowed.

Write your answers

When answering essays under exam conditions it is important that you keep to your time plan as it can be easy to lose track of time if you get into the flow of writing. One way to do this is to make a point of checking the time when you come to the end of a section in your answer. This is preferable to checking the time at random intervals (which could be insufficient) or too frequently (which could distract you from writing fluently). Also, keep looking at the question to remind yourself of the focus of your answer, and keep looking at your plan to remind yourself of what you need to cover. Again, the end of each section of your answer makes a good natural break for checking that you're still on track.

Review your answers

Being able to review your answers is where the benefits of having planned and monitored your use of time become obvious. Not only should you have had enough time to make a reasonable attempt at each of your chosen questions, you should also have time left at the end to go back and review your responses. As we highlighted in Chapter 5, *Writing essays*, making time to review your work can make a significant difference to your final grade.

   As with multiple choice and short answer questions, first check that you have answered every question that you need to answer, then review your answers, starting with the one you were least sure about. The purpose of reviewing answers in an exam context is slightly different to reviewing answers in a coursework setting: you don't need to finish up with a very polished piece of writing which is spelling error-free and grammatically faultless, just one that answers the question well. Of course, if you do end up with a polished piece of writing then that's good, but it's not the primary purpose of a review in an exam context.

   Instead, you should be looking at the bigger issues, such as:

- have you answered the question?
- does the order of your material make sense?
- is the meaning of what you've written clear?
- have you made appropriate use of figures to supplement what you've written?

These are the main issues; you will get more marks for a response that has spelling and grammatical errors but answers the question well than you will for a response that has perfect spelling and grammar but doesn't answer the question well.

If you need to make significant structural changes to your answer, such as moving a paragraph from one section to another, you need to do this in a way that does not confuse the people marking it. In a word-processed script this would be straightforward, but in a hand-written script you will need to annotate it in a way that makes it clear you are moving a section rather than, for instance, deleting it.

✳ Chapter summary

Performing well in exams requires particular skills. Although your exam performance will be heavily influenced by how well you have revised, it is how you conduct yourself during an exam that will ultimately determine your grade. What you do just before an exam and when you arrive in the exam room are important stages of your preparation. Different types of question require slightly different techniques, but all require you to analyse questions, plan answers, write answers, and review answers.

Index

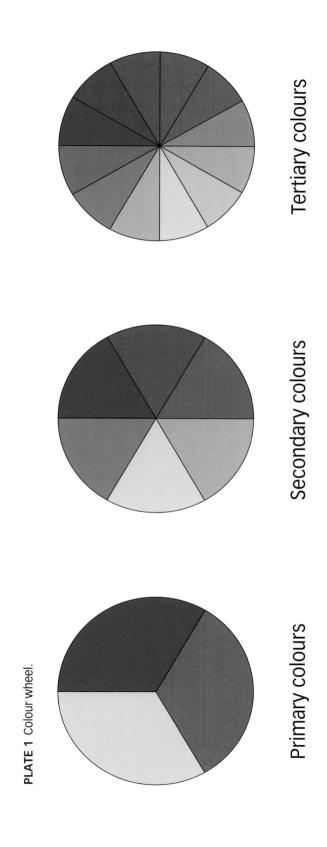

PLATE 1 Colour wheel.

Primary colours

Secondary colours

Tertiary colours